FORCING, ITERATED ULTRAPOWERS, AND TURING DEGREES

LECTURE NOTES SERIES
Institute for Mathematical Sciences, National University of Singapore

Series Editors: Chitat Chong and Wing Keung To
Institute for Mathematical Sciences
National University of Singapore

ISSN: 1793-0758

Published

Vol. 20 Mathematical Horizons for Quantum Physics
edited by H Araki, B-G Englert, L-C Kwek & J Suzuki

Vol. 21 Environmental Hazards:
The Fluid Dynamics and Geophysics of Extreme Events
edited by H K Moffatt & E Shuckburgh

Vol. 22 Multiscale Modeling and Analysis for Materials Simulation
edited by Weizhu Bao & Qiang Du

Vol. 23 Geometry, Topology and Dynamics of Character Varieties
edited by W Goldman, C Series & S P Tan

Vol. 24 Complex Quantum Systems: Analysis of Large Coulomb Systems
edited by Heinz Siedentop

Vol. 25 Infinity and Truth
edited by Chitat Chong, Qi Feng, Theodore A Slaman & W Hugh Woodin

Vol. 26 Notes on Forcing Axioms
by Stevo Todorcevic, edited by Chitat Chong, Qi Feng, Theodore A Slaman, W Hugh Woodin & Yue Yang

Vol. 27 *E*-Recursion, Forcing and C*-Algebras
edited by Chitat Chong, Qi Feng, Theodore A Slaman, W Hugh Woodin & Yue Yang

Vol. 28 Slicing the Turth: On the Computable and Reverse Mathematics of Combinatorial Principles
by Denis R. Hirschfeldt, edited by Chitat Chong, Qi Feng, Theodore A Slaman, W Hugh Woodin & Yue Yang

Vol. 29 Forcing, Iterated Ultrapowers, and Turing Degrees
edited by Chitat Chong, Qi Feng, Theodore A Slaman, W Hugh Woodin & Yue Yang

Vol. 30 Modular Representation Theory of Finite and *p*-Adic Groups
edited by Wee Teck Gan & Kai Meng Tan

*For the complete list of titles in this series, please go to
http://www.worldscientific.com/series/LNIMSNUS

Lecture Notes Series, Institute for Mathematical Sciences,
National University of Singapore
Vol. 29

FORCING, ITERATED ULTRAPOWERS, AND TURING DEGREES

Editors

Chitat Chong
National University of Singapore, Singapore

Qi Feng
Chinese Academy of Sciences, China

Theodore A Slaman
University of California, Berkeley, USA

W Hugh Woodin
Harvard University, USA

Yue Yang
National University of Singapore, Singapore

World Scientific

NEW JERSEY • LONDON • SINGAPORE • BEIJING • SHANGHAI • HONG KONG • TAIPEI • CHENNAI

Published by

World Scientific Publishing Co. Pte. Ltd.
5 Toh Tuck Link, Singapore 596224
USA office: 27 Warren Street, Suite 401-402, Hackensack, NJ 07601
UK office: 57 Shelton Street, Covent Garden, London WC2H 9HE

Library of Congress Cataloging-in-Publication Data
Forcing, iterated ultrapowers, and Turing degrees / edited by Chitat Chong (NUS, Singapore), Qi Feng (Chinese Academy of Sciences, China), Theodore A. Slaman (UC Berkeley), W. Hugh Woodin (Harvard), Yue Yang (NUS, Singapore).
pages cm. -- (Lecture notes series, Institute for Mathematical Sciences, National University of Singapore ; volume 29)
Includes bibliographical references and index.
ISBN 978-9814699945 (hardcover : alk. paper)
1. Model theory. 2. Forcing (Model theory). 3. Logic, Symbolic and mathematical. 4. Unsolvability (Mathematical logic). I. Chong, C.-T. (Chi-Tat), 1949– editor. II. Feng, Qi, 1955– editor. III. Slaman, T. A. (Theodore Allen), 1954– editor. IV. Woodin, W. H. (W. Hugh), editor. V. Yang, Yue, 1964– editor.
QA9.7.F67 2015
511.3'4--dc23

2015018233

British Library Cataloguing-in-Publication Data
A catalogue record for this book is available from the British Library.

Printed in Singapore

CONTENTS

FOREWORD

The Institute for Mathematical Sciences (IMS) at the National University of Singapore was established on 1 July 2000. Its mission is to foster mathematical research, both fundamental and multidisciplinary, particularly research that links mathematics to other efforts of human endeavor, and to nurture the growth of mathematical talent and expertise in research scientists, as well as to serve as a platform for research interaction between scientists in Singapore and the international scientific community.

The Institute organizes thematic programs of longer duration and mathematical activities including workshops and public lectures. The program or workshop themes are selected from among areas at the forefront of current research in the mathematical sciences and their applications.

Each volume of the *IMS Lecture Notes Series* is a compendium of papers based on lectures or tutorials delivered at a program/workshop. It brings to the international research community original results or expository articles on a subject of current interest. These volumes also serve as a record of activities that took place at the IMS.

We hope that through the regular publication of these *Lecture Notes* the Institute will achieve, in part, its objective of reaching out to the community of scholars in the promotion of research in the mathematical sciences.

April 2015

Chitat Chong
Wing Keung To
Series Editors

PREFACE

The series of Asian Initiative for Infinity (AII) Graduate Logic Summer School was held annually from 2010 to 2012. The lecturers were Moti Gitik, Denis Hirschfeldt and Menachem Magidor in 2010, Richard Shore, Theodore A. Slaman, John Steel, and W. Hugh Woodin in 2011, and Ilijas Farah, Ronald Jensen, Gerald E. Sacks and Stevo Todorcevic in 2012. In all, more than 150 graduate students from Asia, Europe and North America attended the summer schools. In addition, two postdoctoral fellows were appointed during each of the three summer schools. These volumes of lecture notes serve as a record of the AII activities that took place during this period.

The AII summer schools were funded by a grant from the John Templeton Foundation and partially supported by the National University of Singapore. Their generosity is gratefully acknowledged.

April 2015

Chitat Chong
National University of Singapore, Singapore

Qi Feng
Chinese Academy of Sciences, China

Theodore A. Slaman
University of California, Berkeley, USA

W. Hugh Woodin
Harvard University, USA

Yue Yang
National University of Singapore, Singapore

Volume Editors

PRIKRY-TYPE FORCINGS AND A FORCING WITH SHORT EXTENDERS

Moti Gitik

School of Mathematical Sciences
Tel Aviv University
Tel Aviv, Israel
gitik@post.tau.ac.il

We cover the following topics: Basic Prikry forcing, tree Prikry forcing, supercompact Prikry forcing, a negation of the Singular Cardinal Hypothesis via blowing up the power of a singular cardinal and short extender forcing — gap 2.

1. Lecture 1 — The Basic Prikry Forcing

Let κ be a measurable cardinal and U a normal ultrafilter over κ.

Definition 1.1 Let $\mathcal{P}$ be the set of all pairs $\langle p, A\rangle$ such that

(a) p is a finite subset of κ,
(b) $A \in U$,
(c) $\min(A) > \max(p)$.

It is convenient sometimes to view p as an increasing finite sequence of ordinals.

We define two partial orderings on $\mathcal{P}$, the first one very unclosed and the second closed enough to compensate the lack of closure of the first.

Definition 1.2 Let $\langle p, A\rangle$ and $\langle q, B\rangle$ be elements of $\mathcal{P}$. We say that $\langle p, A\rangle$ is stronger than $\langle q, B\rangle$ and denote this by $\langle q, B\rangle \leq \langle p, A\rangle$ iff

(1) p is an end extension of q, i.e. $p \cap (\max(q) + 1) = q$,
(2) $A \subseteq B$,
(3) $p \backslash q \subseteq B$.

Definition 1.3 Let $\langle p, A\rangle$ and $\langle q, B\rangle$ be elements of $\mathcal{P}$. We say that $\langle p, A\rangle$ is a direct (or Prikry) extension of $\langle q, B\rangle$ and denote this by $\langle q, B\rangle \leq^* \langle p, A\rangle$ iff

(1) $p = q$,
(2) $A \subseteq B$.

We will force with $\langle \mathcal{P}, \leq \rangle$, and $\langle \mathcal{P}, \leq^* \rangle$ will be used to show that no new bounded subsets are added to κ after the forcing with $\langle \mathcal{P}, \leq \rangle$.

Let us prove a few basic lemmas.

Lemma 1.4 *Let $G \subseteq \mathcal{P}$ be generic for $\langle \mathcal{P}, \leq \rangle$. Then $\bigcup\{p \mid \exists A \in U \langle p, A \rangle \in G\}$ is an ω-sequence cofinal in κ.*

Proof. Just note that for every $\alpha < \kappa$ and $\langle q, B \rangle \in \mathcal{P}$ the set

$$D_\alpha = \{\langle p, A \rangle \in \mathcal{P} \mid \langle p, A \rangle \geq \langle q, B \rangle \quad \text{and} \quad \max(p) > \alpha\}$$

is dense in $\langle \mathcal{P}, \leq \rangle$ above $\langle q, B \rangle$.
□

Lemma 1.5 $\langle \mathcal{P}, \leq \rangle$ *satisfies the κ^+-c.c.*

Proof. Note that any two conditions having the same first coordinate are compatible: If $\langle p, A \rangle$, $\langle p, B \rangle \in \mathcal{P}$, then $\langle p, A \cap B \rangle$ is stronger than both of them.
□

Let us state now three lemmas about $\leq^*$ and its relation to $\leq$. The third one contains the crucial idea of Prikry that makes everything work.

Lemma 1.6 $\leq^* \subseteq \leq$.

This is obvious from Definitions 1.2 and 1.3.

Lemma 1.7 $\langle \mathcal{P}, \leq^* \rangle$ *is κ-closed.*

Proof. Let $\langle \langle p_\alpha, A_\alpha \rangle \mid \alpha < \lambda \rangle$ be a $\leq^*$-increasing sequence of length λ for some $\lambda < \kappa$. Then all the p_α's are the same. Set $p = p_0$ and $A = \bigcap_{\alpha < \kappa} A_\alpha$. Then $A \in U$ by κ-completeness of U. So $\langle p, A \rangle \in \mathcal{P}$, and it is stronger than each $\langle p_\alpha, A_\alpha \rangle$ according to $\leq^*$.
□

Lemma 1.8 *(Prikry condition). Let $\langle q, B \rangle \in \mathcal{P}$ and σ be a statement of the forcing language of $\langle \mathcal{P}, \leq \rangle$. Then there is $\langle p, A \rangle \geq^* \langle q, B \rangle$ such that $\langle p, A \rangle \parallel \sigma$ (i.e. $\langle p, A \rangle \Vdash \sigma$ or $\langle p, A \rangle \Vdash \neg\sigma$), where, again, we force with $\langle \mathcal{P}, \leq \rangle$ and not with $\langle \mathcal{P}, \leq^* \rangle$.*

Proof. We identify finite subsets of κ and finite increasing sequences of ordinals below κ, i.e. $[\kappa]^{<\omega}$. Define a partition $h : [B]^{<\omega} \to 2$ as follows:

$$h(s) = \begin{cases} 1, & \text{if there is a } C \text{ such that } \langle q \cup s, C\rangle \Vdash \sigma \\ 0, & \text{otherwise.} \end{cases}$$

U is a normal ultrafilter, so by the Rowbottom theorem (see, for example, [1, 7.17] or [2, 70]) there is an $A \in U$, $A \subseteq B$ homogeneous for h, i.e. for every $n < \omega$ and every $s_1, s_2 \in [A]^n$, $h(s_1) = h(s_2)$. Now $\langle q, A\rangle$ will decide σ. Since otherwise there will be

$$\langle q \cup s_1, B_1\rangle,\ \langle q \cup s_2, B_2\rangle \geq \langle q, A\rangle$$

such that $\langle q \cup s_1, B_1\rangle \Vdash \sigma$ and $\langle q \cup s_2, B_2\rangle \Vdash \neg\sigma$.

By extending one of these conditions if necessary, we can assume that $|s_1| = |s_2|$. But then $s_1, s_2 \in [A]^{|s_1|}$ and $h(s_1) \neq h(s_2)$, which contradicts the homogeneity of A.

□

The above lemma allows us to implement the κ-closure of $\langle \mathcal{P}, \leq^*\rangle$ in the actual forcing $\langle \mathcal{P}, \leq\rangle$. Thus we can conclude the following:

Lemma 1.9 *$\langle \mathcal{P}, \leq\rangle$ does not add new bounded subsets of κ.*

Proof. Let $t \in \mathcal{P}$, $\underset{\sim}{a}$ is a name, $\lambda < \kappa$ and

$$t \Vdash \underset{\sim}{a} \subseteq \check{\lambda}\ .$$

For every $\alpha < \lambda$ denote by σ_α the statement "$\check{\alpha} \in \underset{\sim}{a}$". We define by recursion a $\leq^*$-increasing sequence of conditions $\langle t_\alpha \mid \alpha < \lambda\rangle$ such that $t_\alpha \parallel \sigma_\alpha$ for each $\alpha < \lambda$. Let t_0 be a direct extension of t deciding σ_0. It exists by 1.8. Suppose that $\langle t_\beta \mid \beta < \alpha\rangle$ is defined. Define t_α. First, using 1.7 we find a direct extension t'_α of $\langle t_\beta \mid \beta < \alpha\rangle$. Then by 1.8 choose a direct extension t_α of t'_α deciding σ_α. This completes the definition of $\langle t_\alpha \mid \alpha < \lambda\rangle$. Now let t^* be a direct extension of $\langle t_\alpha \mid \alpha < \lambda\rangle$ (again 1.7 is used). Then $t^* \geq t$ (in fact $t^* \geq^* t$) and $t^* \Vdash \underset{\sim}{a} = \check{b}$ where $b = \{\alpha < \lambda \mid t^* \Vdash \alpha \in \underset{\sim}{a}\}$.

□

Let us summarize the situation.

Theorem 1.10 *The following holds in $V[G]$:*

(a) *κ has cofinality $\aleph_0$.*
(b) *all the cardinals are preserved.*
(c) *no new bounded subsets are added to κ.*

Proof. (a) is established by 1.4, (c) by 1.9. Finally, (b) follows from (c) and 1.5.
□

If $2^\kappa > \kappa^+$ in V, then in $V[G]$ the Singular Cardinal Hypothesis will fail at κ.

Let $C = \bigcup\{p \mid \exists A \in U \ \langle p, A\rangle \in G\}$. By 1.4, C is an ω-sequence cofinal in κ. It is called a Prikry sequence for U. The generic set G can be easily reconstructed from C:

$$G = \{\langle p, A\rangle \in \mathcal{P} \mid p \text{ is an initial segment of } C \text{ and } C \setminus \max(p) + 1 \subseteq A\} .$$

So, $V[G] = V[C]$.

Lemma 1.11 *C is almost contained in every set in U, i.e.*
(*) for every $A \in U$ the set $C \setminus A$ is finite.

Proof. Let $A \in U$. Then the set

$$D = \{\langle p, B\rangle \in \mathcal{P} \mid B \subseteq A\}$$

is dense in $\mathcal{P}$. So, there is $\langle q, S\rangle \in G \cap D$. But then, for every $\langle q', S'\rangle \geq \langle q, S\rangle q' \setminus q \subseteq S \subseteq A$. Hence, also, $C \setminus q \subseteq A$.
□

The above implies that C generates U, i.e. $X \in U$ iff $X \in V$ and $C \setminus X$ is finite.

A. Mathias pointed out that (*) of 1.11 actually characterizes Prikry sequences:

Theorem 1.12 *Suppose that M is an inner model of ZFC, U a normal ultrafilter over κ in M. Assume that C is an ω-sequence satisfying (*). Then C is a Prikry sequence for U over M.*

Proof. We need to show that the set

$$G(C) = \{\langle p, A\rangle \in \mathcal{P} \mid p \text{ is an initial segment of } C \text{ and } C \setminus \max(p) + 1 \subseteq A\}$$

is a generic subset of $\mathcal{P}$ over M. The only nontrivial property to check is that $G(C) \cap D \neq \emptyset$ for every dense open subset $D \in M$ of $\mathcal{P}$. Let us first point out that the following holds in M:

Lemma 1.13 *Let $\langle q, B\rangle \in \mathcal{P}$ and $D \subseteq \mathcal{P}$ be dense open. Then there are $\langle q, A\rangle \geq^* \langle q, B\rangle$ and $m < \omega$ such that for every n, $m \leq n < \omega$ and every $s \in [A]^n$ we have $\langle q \cup s, A\backslash(\max(s) + 1)\rangle \in D$.*

Proof. We define a partition $h : [B]^{<\omega} \to 2$ as in 1.8 only replacing "$\Vdash \sigma$" by "$\in D$". Let $A' \in U$, $A' \subseteq B$ be homogeneous for h. Then, starting with some m, for every $n \geq m$ and $s \in [A']^n$ we have $h(s) = 1$. Hence there will be a set $A_s \in U$ such that $\langle q \cup s, A_s\rangle \in D$. Set $A = A' \cap \Delta\{A_s \mid s \in [A']^n, m \leq n < \omega\}$, where

$$\Delta\{A_s \mid s \in [A']^n, m \leq n < \omega\}$$

$$= \{\alpha < \kappa \mid \forall n \geq m \forall s \in [A']^n(\max(s) < \alpha \to \alpha \in A_s\} .$$

Then, clearly $A \in U$. The condition $\langle q, A\rangle$ is as desired, since for each $n \geq m$ and $s \in [A]^n$ we have $A\backslash(\max(s)+1) \subseteq A_s$ and, so $\langle q \cup s, A\backslash(\max(s)+1)\rangle \in D$.
□

Now, let $D \in M$ be a dense open subset of $\mathcal{P}$. For every finite $q \subseteq \kappa$, using 1.13, we pick $m(q) < \omega$ and $A(q) \in U$ such that $\langle q, A(q)\rangle \geq^* \langle q, \kappa\backslash \max(q) + 1\rangle$ and for every $n \geq m(q)$ and $s \in [A(q)]^n$ $\langle q \cup s, A(q)\backslash \max(s) + 1\rangle \in D$. Set

$$A = \Delta\{A(q) \mid q \in [\kappa]^{<\omega}\} = \{\alpha < \kappa \mid \forall q \in [\kappa]^{<\omega}(\max q < \alpha \to \alpha \in A(q)\} .$$

There is a $\tau < \kappa$ such that $C\backslash\tau \subseteq A$. Consider $\langle C \cap \tau, A\backslash\tau\rangle$. Since $C \cap \tau$ is finite, $\langle C \cap \tau, A\backslash\tau\rangle \in \mathcal{P}$. Then, for every $n \geq m(C \cap \tau)$ and $s \in [C\backslash\tau]^n$ we have

$$\langle(C \cap \tau) \cup s, A\backslash \max(s) + 1\rangle \in D ,$$

since $A\backslash\tau \subseteq A(C \cap \tau)$. But $C\backslash\tau \subseteq A$, so we can pick $s \in [C\backslash\tau]^n$ for some $n \geq m(C \cap \tau)$. Then $(C \cap \tau) \cup s \subseteq C$ and $C\backslash \max(s) + 1 \subseteq A\backslash \max(s) + 1$. Hence, $\langle(C \cap \tau) \cup s, A\backslash \max(s) + 1\rangle \in G(C) \cap D$. ⊣

1.1. *Lecture 2 — The tree Prikry forcing*

We would now like to eliminate the normality of the ultrafilter U of the previous construction. Note that it was used only once in the proof of the Prikry property 1.8.

Let us now assume only that U is a κ-complete ultrafilter over κ.

Definition 1.14 A set T is called a U-tree with a trunk t iff

(a) T consists of finite increasing sequences of ordinals below κ.
(b) $\langle T, \trianglelefteq\rangle$ is a tree, where $\trianglelefteq$ is the order of end extension of finite sequences, i.e. $\eta \trianglelefteq \nu$ iff $\nu \restriction dom(\eta) = \eta$.
(c) t is a trunk of T, i.e. $t \in T$ and for every $\eta \in T$, $\eta \trianglerighteq t$ or $t \trianglerighteq \eta$.
(d) for every $\eta \trianglerighteq t$ the set $Suc_T(\eta) = \{\alpha < \kappa \mid \eta^\frown\langle\alpha\rangle \in T\}$ is in U.

Define $Lev_n(T) = \{\eta \in T \mid length(\eta) = n\}$ for every $n < \omega$.
Now we define the tree Prikry forcing.

Definition 1.15 The set $\mathcal{P}$ consists of all pairs $\langle t, T\rangle$ such that T is a U-tree with trunk t.

Definition 1.16 Let $\langle t, T\rangle$ and $\langle s, S\rangle$ be in $\mathcal{P}$. We say that $\langle t, T\rangle$ is stronger than $\langle s, S\rangle$ and denote this by $\langle t, T\rangle \geq \langle s, S\rangle$ iff $S \supseteq T$.

Note that $S \supseteq T$ implies that $t \trianglerighteq s$ and $t \in S$.

Definition 1.17 Let $\langle t, T\rangle$ and $\langle s, S\rangle$ be in $\mathcal{P}$. We say that $\langle t, T\rangle$ is a direct (or Prikry) extension of $\langle s, S\rangle$ and denote this by $\langle s, S\rangle \leq^* \langle t, T\rangle$ iff

(1) $S \supseteq T$
(2) $s = t$.

As in the previous section we will force with $\langle \mathcal{P}, \leq\rangle$ and the role of $\leq^*$ will be to provide closure.

Lemma 1.18 *Let $\langle T_\alpha \mid \alpha < \lambda\rangle$ be a sequence of U-trees with the same trunk and $\lambda < \kappa$. Then $T = \bigcap_{\alpha<\lambda} T_\alpha$ is a U-tree having the same trunk as those of the T_α's.*

Proof. Let t be the trunk of T_0 (and so of every T_α). Suppose that $\eta \in T$ and $\eta \trianglerighteq t$. Then

$$Suc_T(\eta) = \bigcap_{\alpha<\lambda} Suc_{T_\alpha}(\eta) .$$

By κ-completeness of U, $Suc_T(\eta) \in U$. Hence T is a U-tree with trunk t.
□

Using 1.18 it is easy to prove lemmas analogous to 1.4–1.7.

Lemma 1.19 *Let $G \subseteq \mathcal{P}$ be generic for $\langle \mathcal{P}, \leq\rangle$. Then*

$$\bigcup\{t \mid \exists T(\langle t, T\rangle \in G)\}$$

is an ω-sequence cofinal in κ.

Lemma 1.20 *$\langle \mathcal{P}, \leq \rangle$ satisfies the κ^+-c.c.*

Lemma 1.21 $\leq^* \subseteq \leq$.

Lemma 1.22 *$\langle \mathcal{P}, \leq^* \rangle$ is κ-closed.*

Let us show that $\langle \mathcal{P}, \leq, \leq^* \rangle$ satisfies the Prikry condition.

The proof is based on the following Ramsey property:

If T is any U-tree and $f : T \to \lambda < \kappa$ then there is an U-tree $S \subseteq T$ such that $f \restriction Lev_n(S)$ is constant for each $n < \omega$.

We prefer here and later to give a direct proof instead of deducing first a relevant Ramsey property and then proving it.

Lemma 1.23 *Let $\langle t, T \rangle \in \mathcal{P}$ and σ be a statement of the forcing language. Then there is a $\langle s, S \rangle \geq^* \langle t, T \rangle$ such that $\langle s, S \rangle \parallel \sigma$.*

Proof. Suppose otherwise. Consider the set $Suc_T(t)$. We split it into three sets as follows:

$$X_0 = \{\alpha \in Suc_T(t) \mid \exists S_\alpha \subseteq T \text{ a } U\text{-tree with trunk } t^\frown\langle\alpha\rangle \text{ such that } \langle t^\frown\langle\alpha\rangle\ ,\ S_\alpha\rangle \Vdash \sigma\}$$

$$X_1 = \{\alpha \in Suc_T(t) \mid \exists S_\alpha \subseteq T \text{ a } U\text{-tree with trunk } t^\frown\langle\alpha\rangle \text{ such that } \langle t^\frown\langle\alpha\rangle, S_\alpha\rangle \Vdash \neg\sigma\}$$

$$X_2 = Suc_T(t) \setminus (X_0 \cup X_1)\ .$$

Clearly $X_0 \cap X_1 = \emptyset$, since by 1.18 any two conditions with the same trunk are compatible. Now U is an ultrafilter and $Suc_T(t) \in U$, so for some $i < 3$ $X_i \in U$. We shrink T to a tree T_1 with the same trunk t, having $Suc_{T_1}(t) = X_i$ and: if $i < 2$ then let T_1 be S_α above $t^\frown\langle\alpha\rangle$ for every $\alpha \in X_i$; if $i = 2$, then let T_1 be the same as T above $t^\frown\langle\alpha\rangle$ for every $\alpha \in X_2$. We continue by recursion to shrink the initial tree T level by level. Thus define a decreasing sequence $\langle T_n \mid n < \omega \rangle$ of U-trees with trunk t so that

(1) $T_0 = T$
(2) for every $n > 0$ for every $m > n$, $T_m \restriction n + |t| = T_n \restriction n + |t|$, i.e. after stage n the n-th level above the trunk remains unchanged in all T_m's for $m \geq n$

(3) for every $n > 0$ if $i < 2$, $\eta \in Lev_{n+|t|}(T_n)$ and for some U-tree S with trunk η

$$\langle \eta, S \rangle \Vdash {}^i\sigma$$

then

(a) $\langle \eta, (T_n)_\eta \rangle \Vdash {}^i\sigma$ and
(b) for every $\nu \in Lev_{n+|t|}(T_n)$ having the same immediate predecessor as η

$$\langle \nu, (T_n)_\nu \rangle \Vdash {}^i\sigma \ .$$

Here, ${}^0\sigma$ denotes σ, ${}^1\sigma$ denotes $\neg\sigma$ and for a tree R with $r \in R$

$$(R)_r = \{r' \in R \mid r' \trianglerighteq r\} \ .$$

Now, we set $T^* = \bigcap_{n<\omega} T_n$. Clearly, T^* is a U-tree with a trunk t by (2) or by 1.18. Consider $\langle t, T^* \rangle \in \mathcal{P}$. By the assumption $\langle t, T^* \rangle \nVdash \sigma$. Pick a condition $\langle s, S \rangle \geq \langle t, T^* \rangle$ forcing σ with $n = |s - t|$ as small as possible. Then $s \in Lev_{n+|t|}(T^*) = Lev_{n+|t|}(T_n)$. By (3) of the recursive construction,

$$\langle s, (T_n)_s \rangle \Vdash \sigma$$

and for every $s' \in Lev_{n+|t|}(T_n)$ with the same predecessor as s, $\langle s', (T_n)_{s'} \rangle \Vdash \sigma$. But $T^* \subseteq T_n$, so

$$\langle s, (T^*)_s \rangle \Vdash \sigma \quad \text{and}$$
$$\langle s', (T^*)_{s'} \rangle \Vdash \sigma$$

for every s' as above.

Let s^* denote the immediate predecessor of s, i.e. s without its last element. Then $\langle s^*, (T^*)_{s^*} \rangle \Vdash \sigma$ since for every $\langle r, R \rangle \geq \langle s^*, (T^*)_{s^*} \rangle$, $r = s'^\frown r'$ for some $s' \in Lev_{n+|t|}(T^*)$ and $s' \triangleright s^*$. Hence, $\langle r, R \rangle \geq \langle s', (T^*)_{s'} \rangle \Vdash \sigma$.

But we chose s to be of the minimal length such that for some $S\langle s, S \rangle \Vdash \sigma$, yet $|s^*| = |s| - 1$. Contradiction.
□

Now, as in 1.9 the κ-closure of $\langle \mathcal{P}, \leq^* \rangle$ can be used to derive the following:

Lemma 1.24 *$\langle \mathcal{P}, \leq \rangle$ does not add new bounded subsets of κ.*

The conclusion is the same as those of the previous section.

Theorem 1.25 *The following holds in $V[G]$:*

(a) *κ has cofinality $\aleph_0$.*
(b) *all the cardinals are preserved.*
(c) *no new bounded subsets are added to κ.*

1.2. *One element Prikry forcing and adding a Prikry sequence to a singular cardinal*

Suppose that κ is a limit of an increasing sequence $\langle \kappa_n \mid n < \omega \rangle$ of measurable cardinals. We want to add an ω-sequence dominating every sequence in $\prod_{n<\omega} \kappa_n$, i.e. a sequence $\langle \tau_m \mid m < \omega \rangle \in \prod_{n<\omega} \kappa_n$ such that for every $\langle \rho_m \mid m < \omega \rangle \in \left(\prod_{n<\omega} \kappa_n \right) \cap V$ and for all but finitely many m's, $\tau_m > \rho_m$.

Fix a κ_n-complete ultrafilter U_n over κ_n for every $n < \omega$. One can assume normality but it is not necessary.

Let $n < \omega$. We describe first a very simple forcing for adding a one element Prikry sequence.

Definition 1.26 Let $Q_n = U_n \cup \kappa_n$. If $p, q \in Q_n$ we define $p \geq_n q$ iff either

(i) $p, q \in U_n$ and $p \subseteq q$.
(ii) $q \in U_n$ and $p \in q$.
(iii) $p = q \in \kappa_n$.

Thus we can pick a set in U_n, and then shrink it still in U_n or pick an element of this set. In particular, above every condition there is an atomic one. So the forcing $\langle Q_n, \leq_n \rangle$ is trivial.

Nevertheless we also define a direct extension ordering:

Definition 1.27 Let $p, q \in Q_n$. Set $p \geq_n^* q$ iff $p = q$ or $(p, q \in U_n$ and $p \subseteq q)$.

The forcing $\langle Q_n, \leq_n, \leq_n^* \rangle$ is called the one element Prikry forcing. The following lemma follows from the κ_n-completeness of U_n.

Lemma 1.28 *$\langle Q_n, \leq_n^* \rangle$ is κ_n-closed.*

Lemma 1.29 *$\langle Q_n, \leq_n, \leq_n^* \rangle$ satisfies the Prikry condition, i.e. for every $p \in Q_n$ and every statement σ of the forcing language there is a $q \geq_n^* p$ such that $q \parallel \sigma$.*

The proof repeats the first stage of the proof of 1.23.

Now we combine Q_n's together.

Definition 1.30 Let $\mathcal{P}$ be the set of all sequences $p = \langle p_n \mid n < \omega \rangle$ so that

(1) for every $n < \omega$, $p_n \in Q_n$
(2) there is $\ell(p) < \omega$ so that for every $n < \ell(p)$, p_n is an ordinal below κ_n and for every $n \geq \ell(p)$, $p_n \in U_n$.

The orderings $\leq$ and $\leq^*$ are defined on $\mathcal{P}$ in obvious fashion:

Definition 1.31 Let $p = \langle p_n \mid n < \omega \rangle$ and $q = \langle q_n \mid n < \omega \rangle \in \mathcal{P}$. We define $p \geq q$ ($p \geq^* q$) iff for every $n < \omega$, $p_n \geq_n q_n$ ($p_n \geq^*_n q_n$).

For $p = \langle p_n \mid n < \omega \rangle \in \mathcal{P}$ we denote $\langle p_m \mid m < n \rangle$ by $p \restriction n$ and $\langle p_m \mid m \geq n \rangle$ by $p \backslash n$. Let $\mathcal{P} \restriction n = \{p \restriction n \mid p \in \mathcal{P}\}$ and $\mathcal{P} \backslash n = \{p \backslash n \mid p \in \mathcal{P}\}$.

The following splitting lemma is obvious:

Lemma 1.32 $\mathcal{P} \simeq \mathcal{P} \restriction n \times \mathcal{P} \backslash n$ *for every* $n < \omega$.

Lemma 1.33 *For every* $n < \omega$ $\langle \mathcal{P} \backslash n, \leq^* \rangle$ *is* κ_n*-closed.*

The above follows from the fact that each U_m with $m \geq n$ is κ_n-complete.

Lemma 1.34 $\langle \mathcal{P}, \leq, \leq^* \rangle$ *satisfies the Prikry condition.*

Proof. Let $p = \langle p_n \mid n < \omega \rangle$ be an element of $\mathcal{P}$ and σ be a statement of the forcing language. Suppose for simplicity that $\ell(p) = 0$. Then let $p_n = A_n \in U_n$ for every $n < \omega$. We want to find a direct extension of p deciding σ. Assume that there is no such extension. Define by induction on $n < \omega$ a $\leq^*$-increasing sequence $\langle q(n) \mid n < \omega \rangle$ of $\leq^*$-extensions of p such that for every $n < \omega$ the following holds:

(1) if $m \geq n$, then $q(m) \restriction n = q(n) \restriction n$,
(2) if $q = \langle q_n \mid n < \omega \rangle \geq q(n)$ decides σ and $\ell(q) = n+1$ then already $\langle q_m \mid m \leq n \rangle ^\frown \langle q(n)_m \mid m > n \rangle$ decides σ and in the same way as q; moreover for every $\tau_n \in q(n)_n$ also $\langle q_m \mid m < n \rangle ^\frown \langle \tau_n \rangle ^\frown \langle q(n)_m \mid m > n \rangle$ makes the same decision.

The recursive construction is straightforward. Only at stage n the κ_n-completeness of U_m's ($m \geq n$) is used in order to take care of the possibilities for initial sequences of length $n-1$ below κ_n. The number of such possibilities is $|\prod_{i \leq n-1} \kappa_i| = \kappa_{n-1} < \kappa_n$. Now define $s = \langle s_n \mid n < \omega \rangle$ to be $\langle q(n)_n \mid n < \omega \rangle$. Clearly, $s \in \mathcal{P}$ and $s \geq^* p$. The conclusion is now as in 1.23. Thus let $q = \langle q_n \mid n < \omega \rangle$ be an extension of s forcing σ and with $\ell(q)$

as small as possible. By the assumption, $\ell(q) > 0$. Let $n = \ell(q) - 1$. Now, using (2) of the inductive construction, we conclude that

$$\langle q_m \mid m < n\rangle^\frown\langle \tau_n\rangle^\frown\langle s_m \mid m > n\rangle \Vdash \sigma$$

for every $\tau_n \in q(n)_n = s_n$. But then also $\langle q_m \mid m < n\rangle^\frown\langle s_m \mid m \geq n\rangle \Vdash \sigma$, contradicting the minimality of $\ell(q)$.
□

Combining 1.32, 1.33 and 1.34 we obtain the following:

Lemma 1.35 *$\langle \mathcal{P}, \leq\rangle$ does not add new bounded subsets to κ.*

Note that for each $n < \omega$ $\mathcal{P} \restriction n$ is just a trivial forcing "adding" a sequence of the length n of ordinals in $\prod_{m\leq n-1} \kappa_m$.

Lemma 1.36 *$\langle \mathcal{P}, \leq\rangle$ satisfies the κ^+-c.c.*

Proof. Note that any two conditions $p = \langle p_n \mid n < \omega\rangle$ and $q = \langle q_n \mid n < \omega\rangle$ are compatible provided $\ell(p) = \ell(q)$ and $\langle p_n \mid n < \ell(p)\rangle = \langle q_n \mid n < \ell(q)\rangle$.
□

Now let $G \subseteq \mathcal{P}$ be generic for $\langle \mathcal{P}, \leq\rangle$. Define an ω-sequence $\langle t_n \mid n < \omega\rangle \in \prod_{n<\omega} \kappa_n$ as follows: $t_n = \tau$ if for some $p = \langle p_m \mid m < \omega\rangle \in G$ with $\ell(p) > n$ $p_n = \tau$.

Using density arguments it is easy to show the following:

Lemma 1.37 *For every $\langle s_n \mid n < \omega\rangle \in \left(\prod_{n<\omega} \kappa_n\right) \cap V$ there is $n_0 < \omega$ such that for every $n \geq n_0$, $t_n > s_n$.*

Now combining lemmas together we obtain the following:

Theorem 1.38 *The following holds in $V[G]$:*

(a) all cardinals and cofinalities are preserved.
(b) no new bounded subsets are added to κ.
(c) there is a sequence in $\prod_{n<\omega} \kappa_n$ dominating every sequence in $\left(\prod_{n<\omega} \kappa_n\right) \cap V$.

1.3. *Supercompact and strongly compact Prikry forcings*

We turn now to Prikry forcings that correspond to supercompact and strongly compact cardinals. The main feature of these forcings is that not only κ changes its cofinality to ω, but also every regular cardinal in the interval $[\kappa, \lambda]$ does so, if we use a λ-supercompact (or strongly compact) cardinal κ.

Fix cardinals $\kappa \leq \lambda$. Let $\mathcal{P}_\kappa(\lambda) = \{P \subseteq \lambda \mid |P| < \kappa\}$. Let us recall few basic definitions.

Definition 1.39 An ultrafilter U over $\mathcal{P}_\kappa(\lambda)$ is called normal iff

(a) U is κ-complete.
(b) U is fine, i.e. for every $\alpha < \lambda$, $\{P \in \mathcal{P}_\kappa(\lambda) \mid \alpha \in P\} \in U$.
(c) for every $A \in U$ and every $f : A \to \lambda$ satisfying $f(P) \in P$ for $P \in A$ there are $A' \in U$ and $\alpha' < \lambda$ such that for every $P \in A'$ we have $f(P) = \alpha'$.

Definition 1.40 (1) κ is called λ-strongly compact iff there exists a κ-complete fine ultrafilter over $\mathcal{P}_\kappa(\lambda)$.

(2) κ is called λ-supercompact iff there exists a normal ultrafilter over $\mathcal{P}_\kappa(\lambda)$.

Suppose first that κ is λ-supercompact cardinal. Let U be a normal ultrafilter over $\mathcal{P}_\kappa(\lambda)$.

If $P, Q \in \mathcal{P}_\kappa(\lambda)$, then P is *strongly included* in Q if $P \subseteq Q$ and $otp(P) < otp(Q \cap \kappa)$. Denote this by $P \underset{\sim}{\subset} Q$.

The normality of U easily implies the following:

(A) If F is function from a set in U into $\mathcal{P}_\kappa(\lambda)$ such that for all $P \neq \emptyset$ $F(P) \underset{\sim}{\subset} P$ then F is constant on a set in U.

(B) If for every $Q \in \mathcal{P}_\kappa(\lambda)$, $A_Q \in U$ then $\{P \mid \forall Q \underset{\sim}{\subset} P\ (P \in A_Q)\} \in U$. (This last set is called the diagonal intersection of the system $\{A_Q \mid Q \in \mathcal{P}_\kappa(\lambda)\}$.)

For $B \subseteq \mathcal{P}_\kappa(\lambda)$, denote by $[B]^{[n]}$ the set of all n element subsets of B totally ordered by $\underset{\sim}{\subset}$; denote $\bigcup_{n<\omega}[B]^{[n]}$ by $[B]^{[<\omega]}$. The following is a straightforward analog of the Rowbottom theorem:

If $F : [\mathcal{P}_\kappa(\lambda)]^{[<\omega]} \to 2$ then there is $A \in U$ such that for every $n < \omega$, F is constant on $[A]^{[n]}$.

Now we are ready to define the supercompact Prikry forcing with a normal ultrafilter U over $\mathcal{P}_\kappa(\lambda)$.

The definitions will be the same as in 1.1 with only κ replaced by $\mathcal{P}_\kappa(\lambda)$ and the order on ordinals replaced by "$\underset{\sim}{\subset}$".

Definition 1.41 Let $\mathcal{P}$ be the set of all pairs $\langle\langle P_1, \dots, P_n\rangle, A\rangle$,

(a) $\langle P_1, \dots, P_n\rangle$ is a finite $\underset{\sim}{\subset}$-increasing sequence of elements of $\mathcal{P}_\kappa(\lambda)$,
(b) $A \in U$,
(c) for every $Q \in A$, $Q \underset{\sim}{\supset} P_n$.

Definition 1.42 Let $\langle\langle P_1, \dots, P_n\rangle, A\rangle$ and $\langle\langle Q_1, \dots, Q_m\rangle, B\rangle$ be elements of $\mathcal{P}$.
Then $\langle\langle P_1, \dots, P_n\rangle, A\rangle \geq \langle\langle Q_1, \dots, Q_m\rangle, B\rangle$ iff

(1) $n \geq m$,
(2) for every $k \leq m$ $P_k = Q_k$,
(3) $A \subseteq B$,
(4) $\{P_{m+1}, \dots, P_n\} \subseteq B$.

Definition 1.43 Let $\langle\langle P_1, \dots, P_n\rangle, A\rangle$ and $\langle\langle Q_1, \dots, Q_m\rangle, B\rangle$ be elements of $\mathcal{P}$.
Then $\langle\langle P_1, \dots, P_n\rangle, A\rangle \geq^* \langle\langle Q, \dots, Q_m\rangle, B\rangle$ iff

(1) $\langle P_1, \dots, P_n\rangle = \langle Q_1, \dots, Q_m\rangle$,
(2) $A \subseteq B$.

The next lemmas are proved as in 1.1 with obvious changes from κ to $\mathcal{P}_\kappa(\lambda)$.

Lemma 1.44 $\leq^* \subseteq \leq$.

Lemma 1.45 *$\langle \mathcal{P}, \leq^*\rangle$ is κ-closed.*

Lemma 1.46 *(Prikry condition) Let $\langle q, B\rangle \in \mathcal{P}$ and σ be a statement of the forcing language (i.e. of $\langle \mathcal{P}, \leq\rangle$). Then there is $\langle p, A\rangle \geq^* \langle q, B\rangle$ such that $\langle p, A\rangle \parallel \sigma$.*

Lemma 1.47 *$\langle \mathcal{P}, \leq\rangle$ does not add new bounded subsets to κ.*

Lemma 1.48 *$\langle \mathcal{P}, \leq\rangle$ satisfies the $(\lambda^{<\kappa})^+$-c.c.*

Proof. As in 1.5, any two conditions with the same finite sequence, i.e. of the form $\langle p, A\rangle$ and $\langle p, B\rangle$ are compatible. The number of possibilities for p's now is $\lambda^{<\kappa}$. So we are done.
□

By the theorem of Solovay we have $\lambda^{<\kappa} = \lambda$ if λ is regular or of cofinality $\geq \kappa$ and $\lambda^{<\kappa} = \lambda^+$ in case of $cf(\lambda) < \kappa$. Note that λ-supercompactness of κ implies actually its $\lambda^{<\kappa}$-supercompactness. We can restate 1.48 using Solovay's theorem as follows:

Lemma 1.49 *$\langle \mathcal{P}, \leq\rangle$ satisfies the μ^+-c.c., where*

$$\mu = \begin{cases} \lambda, & \text{if } cf(\lambda) \geq \kappa \\ \lambda^+, & \text{if } cf(\lambda) < \kappa. \end{cases}$$

Our next lemma presents the main property of the supercompact Prikry forcing. Also, it shows that 1.49 is sharp.

Let G be a generic subset of $\langle \mathcal{P}, \leq\rangle$ and let $\langle P_n \mid 1 \leq n < \omega\rangle$ be the Prikry sequence produced by G, i.e. the sequence such that for every $n < \omega$, there is $A \in U$ with $\langle\langle P_1, \ldots, P_n\rangle, A\rangle \in G$.

Lemma 1.50 *Every $\delta \in [\kappa, \mu]$ of cofinality $\geq \kappa$ (in V) changes its cofinality to ω in $V[G]$, where*

$$\mu = \begin{cases} \lambda, & \text{if } cf(\lambda) \geq \kappa \\ \lambda^+, & \text{if } cf(\lambda) < \kappa. \end{cases}$$

Moreover, for each $\delta \leq \lambda$, $\delta = \bigcup_{n<\omega}(P_n \cap \delta)$, i.e. it is a countable union of old sets each of cardinality less than κ.

Proof. Let $\alpha < \lambda$. The fineness of U implies that $\{P \in \mathcal{P}_\kappa(\lambda) \mid \alpha \in P\} \in U$. Then, by a density argument, $\alpha \in P_n$ for all but finitely many n's. Hence, for each $\delta \leq \lambda$

$$\delta = \bigcup_{n<\omega} (P_n \cap \delta).$$

This implies that each $\delta \leq \lambda$ of cofinality $\geq \kappa$ in V changes cofinality to ω in $V[G]$, as witnessed by $\langle \sup(P_n \cap \delta) \mid n < \omega\rangle$. In order to finish the proof, we need to deal with λ of cofinality below κ and to show that in this case λ^+ also changes its cofinality to ω. Fix in V a sequence cofinal in λ of regular cardinals $\langle \lambda_i \mid i < cf(\lambda)\rangle$, a sequence of functions $\langle f_\alpha \mid \alpha < \lambda^+\rangle$ in $\prod_{i<cf\lambda} \lambda_i$ and an ultrafilter D over $cf(\lambda)$ including all cobounded subsets of $cf(\lambda)$, so that

(a) $\alpha < \beta < \lambda^+ \Longrightarrow f_\alpha < f_\beta \pmod D$

(b) for every $g \in \prod_{i<cf\lambda} \lambda_i$ there is $i < \lambda^+$ such that $f_i > g \pmod D$.

Using $\lambda^{<\kappa} = \lambda^+$, it is not hard directly by induction to construct such sequence of f_i's. One can also appeal to general *pcf* considerations. Now, by fineness and density again, for every $\alpha < \lambda^+$ and for all but finitely many $n < \omega$ we will have $P_n \supseteq \mathrm{rng}(f_\alpha)$. Hence, for such n's, $\langle \cup(P_n \cap \lambda_i) \mid i < cf(\lambda)\rangle > f_\alpha$. So, $\{\langle \cup(P_n \cap \lambda_i) \mid i < cf(\lambda)\rangle \mid n < \omega\}$ will be an ω-sequence of functions from $\left(\prod_{i<cf\lambda} \lambda_i\right) \cap V$ unbounded in $\left(\prod_{i<cf\lambda} \lambda_i\right) \cap V$. This implies that λ^+ should have cofinality ω in $V[G]$.
□

Let us now turn to the strongly compact Prikry forcing in 1.2. So, we give up normality and assume only that U is a κ-complete fine ultrafilter over $\mathcal{P}_\kappa(\lambda)$. The construction here is completely parallel to the construction of the tree Prikry forcing in 1.2.

Definition 1.51 A set T is called a U-tree with trunk t iff

(a) T consists of finite sequences $\langle P_1, \ldots, P_n\rangle$ of elements of $\mathcal{P}_\kappa(\lambda)$ so that $P_1 \underset{\sim}{\subset} P_2 \underset{\sim}{\subset} \cdots \underset{\sim}{\subset} P_n$,

(b) $\langle T, \trianglelefteq\rangle$ is a tree, where $\trianglelefteq$ is the order of the end extension of finite sequences,

(c) t is a trunk of T, i.e. $t \in T$ and for every $\eta \in T$, $\eta \trianglerighteq t$ or $t \trianglerighteq \eta$,

(d) for every $\eta \trianglerighteq t$ the set

$$Suc_T(\eta) = \{Q \in \mathcal{P}_\kappa(\lambda) \mid \eta^\frown\langle Q\rangle \in T\}$$

is in U.

The definitions of the forcing notion $\mathcal{P}$ and the orders $\leq$ and $\leq^*$ are now exactly the same as those in 1.15, 1.16 and 1.17. $\langle \mathcal{P}, \leq, \leq^*\rangle$ here shares all the properties of the tree Prikry forcing of Sec. 1.2 except the κ^+-c.c. Thus the Lemmas 1.18, 1.21–1.24 are valid in the present context with basically the same proofs. Instead of the κ^+–c.c. we will have here $(\lambda^{<\kappa})^+$–c.c. Also 1.48–1.50 holds with the same proofs.

Let us summarize the properties of both supercompact and strongly compact Prikry forcings.

Theorem 1.52 *Let G be a generic set for $\langle \mathcal{P}, \leq, \leq^*\rangle$, where $\langle \mathcal{P}, \leq, \leq^*\rangle$ is either supercompact or strongly compact Prikry forcing over $\mathcal{P}_\kappa(\lambda)$. The following holds in $V[G]$:*

(a) no new bounded subsets are added to κ,

(b) every cardinal in the interval $[\kappa, \mu]$ *of cofinality* $\geq \kappa$ *(as computed in V) changes its cofinality to* ω,

(c) all the cardinals above μ *are preserved, where*

$$\mu = \begin{cases} \lambda, & \text{if } cf(\lambda) \geq \kappa \\ \lambda^+, & \text{if } cf(\lambda) < \kappa. \end{cases}$$

2. Lecture 3 — Adding Many Prikry Sequences to a Singular Cardinal

We would like now to present the extender based Prikry forcing over a singular cardinal. It is probably the simplest direct way for violating of the Singular Cardinal Hypothesis.

Let, as in Sec. 1.2, $\kappa = \bigcup_{n<\omega} \kappa_n$ with $\langle \kappa_n \mid n < \omega \rangle$ increasing and each κ_n measurable. The Prikry forcing described in Sec. 1.2 produces basically one Prikry sequence. More precisely, if GCH holds in the ground model, then κ^+-many new ω-sequences are introduced but all of them are coded by the generic Prikry sequence. Here we like to present a way for adding any number of Prikry sequences into $\prod_{n<\omega} \kappa_n$. In particular this will blow the power of κ as large as one likes without adding new bounded subsets and preserving all the cofinalities.

The basic idea is to use many ultrafilters over each of κ_n's instead of a single one used in 1.3. This leads naturally to extenders over κ_n's.

Assume GCH and let $\lambda \geq \kappa^+$ be a regular cardinal. Suppose that we like to add to κ or into $\prod_{n<\omega} \kappa_n$ at least λ many Prikry sequences. Our basic assumption will be now that each κ_n is a $\lambda + 1$-strong cardinal. This means that for every $n < \omega$ there is a $(\kappa_n, \lambda + 1)$-extender E_n over κ_n whose ultrapower contains $V_{\lambda+1}$ and which moves κ_n above λ. We fix such E_n and let $j_n : V \to M_n \simeq Ult(V, E_n)$. For every $\alpha < \lambda$ we define a κ_n-complete ultrafilter $U_{n\alpha}$ over κ_n by setting $X \in U_{n\alpha}$ iff $\alpha \in j_n(X)$. Actually only $U_{n\alpha}$'s with $\alpha \geq \kappa_n$ will be important. Note that a lot of $U_{n\alpha}$'s are comparable in the Rudin-Keisler order $\leq_{RK}$, recalling that $U \leq_{RK} W$ iff there is $f : \cup W \to \cup U$ such that $X \in U$ iff $f^{-1}(X) \in W$. Thus, for example, if α is a cardinal and $\beta \leq \alpha$, then $U_{n,\alpha+\beta} \geq_{RK} U_{n,\alpha}$ and $U_{n,\alpha+\beta} \geq_{RK} U_{n,\beta}$.

We will need a strengthening of the Rudin-Keisler order. Thus, for $\alpha, \beta < \lambda$ let $\alpha \leq_{E_n} \beta$ iff $\alpha \leq \beta$ and for some $f \in {}^{\kappa_n}\kappa_n$, $j_n(f)(\beta) = \alpha$. Clearly, then $\alpha \leq_{E_n} \beta$ implies $U_{n\alpha} \leq_{RK} U_{n\beta}$, as witnessed by any $f \in {}^{\kappa_n}\kappa_n$ with $j_n(f)(\beta) = \alpha$: If $A \in U_{n\beta}$, then $\beta \in j_n(A)$. So $\alpha = j_n(f)(\beta) \in j_n(f)``j_n(A) = j_n(f``(A))$. Hence $f``(A) \in U_{n,\alpha}$. Note that, in general, $\alpha < \beta < \lambda$ and $U_{n\alpha} <_{RK} U_{n\beta}$ does not imply $\alpha <_{E_n} \beta$.

The partial order $\langle \lambda, \leq_{E_n} \rangle$ is κ_n-directed, as we see in Lemma 2.1 below. Actually, it is κ_n^{++}-directed, but for our purposes κ_n-directness will suffice. Thus, using GCH, find some enumeration $\langle a_\alpha \mid \alpha < \kappa_n \rangle$ of $[\kappa_n]^{<\kappa_n}$ so that for every regular cardinal $\delta < \kappa_n$ $\langle a_\alpha \mid \alpha < \delta \rangle$ enumerates $[\delta]^{<\delta}$ and every element of $[\delta]^{<\delta}$ appears δ many times in the enumeration. Let $j_n(\langle a_\alpha \mid \alpha < \kappa_n \rangle) = \langle a_\alpha \mid \alpha < j_n(\kappa_n) \rangle$. Then, $\langle a_\alpha \mid \alpha < \lambda \rangle$ will enumerate $[\lambda]^{<\lambda} \supseteq [\lambda]^{<\kappa_n}$ in both M_n and V; this coding will be applied below.

The next lemma is a basic application of commutativity of diagrams corresponding to extenders and their ultrafilters.

Lemma 2.1 *Let $n < \omega$ and $\tau < \kappa_n$. Suppose that $\langle \alpha_\nu \mid \nu < \tau \rangle$ is a sequence of ordinals below λ and $\alpha \in \lambda \setminus \left(\bigcup_{\nu<\tau} \alpha_\nu + 1 \right)$ codes this sequence, i.e. $a_\alpha = \{\alpha_\nu \mid \nu < \tau\}$. Then $\alpha >_{E_n} \alpha_\nu$ for every $\nu < \tau$.*

Proof. Fix $\nu < \tau$. Consider the following diagram

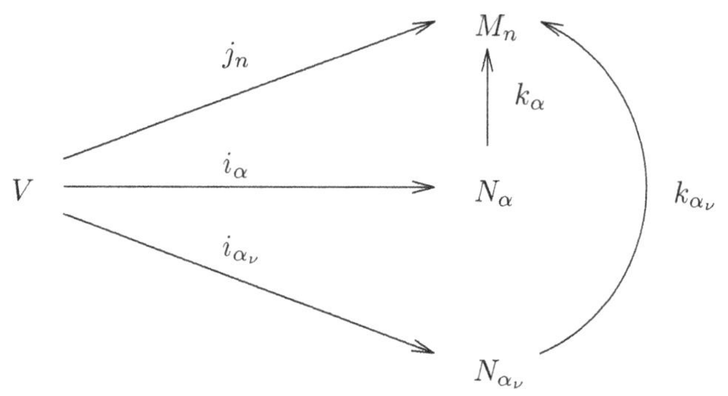

where $N_\alpha \simeq Ult(V, U_{n,\alpha})$, $k_\alpha([f]_{U_{n,\alpha}}) = j_n(f)(\alpha)$ and the same with α_ν replacing α. Then $j_n(\langle a_\beta \mid \beta < \kappa_n\rangle) = k_\alpha(i_\alpha(\langle a_\beta \mid \beta < \kappa_n\rangle))$ and $k_\alpha(i_\alpha(\langle a_\beta \mid \beta < \kappa_n\rangle)([id])_{U_{n,\alpha}})) = j_n(\langle a_\beta \mid \beta < \kappa_n\rangle)(\alpha) = a_\alpha = \{\alpha_\mu \mid \mu < \tau\}$. But $\tau < \kappa_n$, so it is fixed by k_α, since $\mathrm{crit}(k_\alpha) \geq \kappa_n$. Hence $i_\alpha(\langle a_\beta \mid \beta < \kappa_n\rangle)([id]_{U_{n,\alpha}})$ is a sequence of ordinals of length τ. Let α^*_ν denote its ν-th element. Then, by elementarity, $k_\alpha(\alpha^*_\nu) = \alpha_\nu$. We can hence define $k_{\alpha_\nu \alpha} : N_{\alpha_\nu} \to N_\alpha$ by setting $k_{\alpha_\nu\alpha}([f]_{U_{\alpha_\nu}}) = i_\alpha(f)(\alpha^*_\nu)$. It is easy to see that $k_{\alpha_\nu\alpha}$ is elementary embedding and the following diagram is commutative:

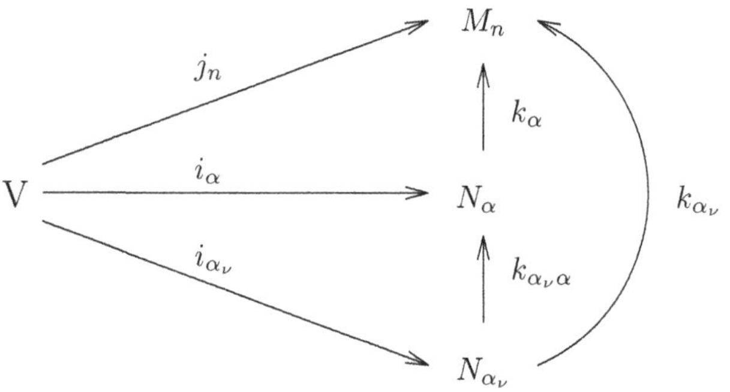

Finally we can define the desired projection $\pi_{\alpha\alpha_\nu}$ of $U_{n,\alpha}$ onto U_{n,α_ν}. Thus let $\pi_{\alpha\alpha_\nu} : \kappa_n \to \kappa_n$ be a function such that $[\pi_{\alpha\alpha_\nu}]_{U_{n,\alpha}} = \alpha^*_\nu$. Then, $j_n(\pi_{\alpha\alpha_\nu})(\alpha) = k_\alpha([\pi_{\alpha,\alpha_\nu}]_{U_{n,\alpha}}) = k_\alpha(\alpha^*_\nu) = \alpha_\nu$. So, $\alpha >_{E_n} \alpha_\nu$.
□

Hence we obtain the following:

Lemma 2.2 *For every set $a \subseteq \lambda$ of cardinality less than κ_n there are λ many α's below λ so that $\alpha >_{E_n} \beta$ for every $\beta \in a$.*

For every $\alpha, \beta < \lambda$ such that $\alpha >_{E_n} \beta$ we fix the projection $\pi_{\alpha\beta} : \kappa_n \to \kappa_n$ defined as in 2.1 witnessing this. Let $\pi_{\alpha\alpha} = id$, the identity map: $\kappa_n \to \kappa_n$.

The following two lemmas are standard.

Lemma 2.3 *Let $\gamma < \beta \leq \alpha < \lambda$. If $\alpha \geq_{E_n} \beta$ and $\alpha \geq_{E_n} \gamma$ then $\{\nu < \kappa_n \mid \pi_{\alpha\beta}(\nu) > \pi_{\alpha\gamma}(\nu)\} \in U_{n\alpha}$.*

Proof. We consider the following commutative diagram:

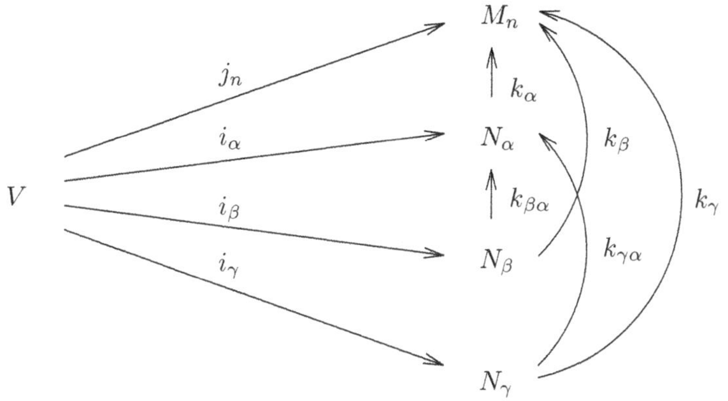

where for $\delta', \delta \in \{\alpha, \beta, \gamma\}$

$$i_\delta : V \to N_\delta \simeq Ult(V, U_{n\delta})$$

$$k_\delta([f]_{U_{n\delta}}) = j_n(f)(\delta)$$

and $k_{\delta'\delta}([f]_{U_{n\delta'}}) = i_\delta(f)([\pi_{\delta\delta'}]_{U_{n\delta}})$.

Then $k_\alpha([\pi_{\alpha\beta}]_{U_{n\alpha}}) = k_\alpha(k_{\beta\alpha}([id]_{U_{n\beta}})) = k_\beta([id]_{U_{n\beta}}) = j_n(id)(\beta) = \beta$. The same is true for γ, i.e.

$$k_\alpha([\pi_{\alpha\gamma}]_{U_{n\alpha}}) = \gamma.$$

But $M_n \vDash \gamma < \beta$ and k_α is elementary, so $N_\alpha \vDash [\pi_{\alpha\gamma}]_{U_{n\alpha}} < [\pi_{\alpha\beta}]_{U_{n\alpha}}$. Hence

$$\{\nu < \kappa_n \mid \pi_{\alpha\beta}(\nu) > \pi_{\alpha\gamma}(\nu)\} \in U_{n\alpha}.$$

□

Lemma 2.4 *Let $\{\alpha_i \mid i < \tau\} \subseteq \alpha < \lambda$ for some $\tau < \kappa_n$. Assume that $\alpha \geq_{E_n} \alpha_i$ for every $i < \tau$. Then there is a set $A \in U_{n\alpha}$ so that for every $i, j < \tau$: $\alpha_i \geq_{E_n} \alpha_j$ implies $\pi_{\alpha\alpha_j}(\nu) = \pi_{\alpha_i\alpha_j}(\pi_{\alpha\alpha_i}(\nu))$ for every $\nu \in A$.*

Proof. It is enough to prove the lemma for $\tau = 2$ and then to use the κ_n-completeness of $U_{n\alpha}$. So, let $\beta, \gamma < \alpha$ and assume that $\gamma \leq_{E_n} \beta \leq_{E_n} \alpha$. Consider the following commutative diagram:

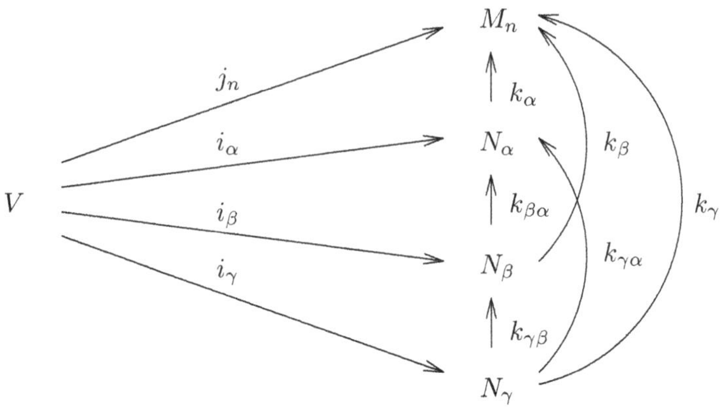

where k's and i's are defined as in 2.3.

We need to show that

$$[\pi_{\alpha\gamma}]_{U_{n\alpha}} = [\pi_{\beta\gamma} \circ \pi_{\alpha\beta}]_{U_{n\alpha}}.$$

As in 2.3, $k_\alpha([\pi_{\alpha\gamma}]_{U_{n\alpha}}) = \gamma$. On the other hand, again as 2.3,

$$\begin{aligned} k_\alpha([\pi_{\beta\gamma} \circ \pi_{\alpha\beta}]_{U_{n\alpha}}) &= j_n(\pi_{\beta\gamma} \circ \pi_{\alpha\beta})(\alpha) \\ &= j_n(\pi_{\beta\gamma})(j_n(\pi_{\alpha\beta})(\alpha)) = j_n(\pi_{\beta\gamma})(\beta) = \gamma. \end{aligned}$$

Since k_α is elementary, we have in N_α the desired equality.
□

We are now ready to define our first forcing notion. It will resemble the one element Prikry forcing considered in 1.3 and will be built from two pieces. Fix $n < \omega$.

Definition 2.5 Let $Q_{n1} = \{f \mid f$ is a partial function from λ to κ_n of cardinality at most $\kappa\}$. We order Q_{n1} by inclusion. Denote this order by $\leq_1$.

Thus Q_{n1} is basically the usual Cohen forcing for blowing up the power of κ^+ to λ. The only minor change is that the functions take values inside κ_n rather than 2 or κ^+.

Definition 2.6 Let Q_{n0} be the set of triples $\langle a, A, f\rangle$ so that

(1) $f \in Q_{n1}$
(2) $a \subseteq \lambda$ such that

  (i) $|a| < \kappa_n$
  (ii) $a \cap \mathrm{dom}(f) = \emptyset$

(iii) a has a $\leq_{E_n}$-maximal element, i.e. an element $\alpha \in a$ such that $\alpha \geq_{E_n} \beta$ for every $\beta \in a$

(3) $A \in U_{n\,\max(a)}$

(4) for every $\alpha, \beta, \gamma \in a$, if $\alpha \geq_{E_n} \beta \geq_{E_n} \gamma$, then $\pi_{\alpha\gamma}(\rho) = \pi_{\beta\gamma}(\pi_{\alpha\beta}(\rho))$ for every $\rho \in \pi_{\max(a),\alpha}$"$A$

(5) for every $\alpha > \beta$ in a and every $\nu \in A$

$$\pi_{\max(a),\alpha}(\nu) > \pi_{\max(a),\beta}(\nu) .$$

The last two conditions can be met by Lemmas 2.3 and 2.4.

Definition 2.7 Let $\langle a, A, f\rangle, \langle b, B, g\rangle \in Q_{n0}$. Then

$$\langle a, A, f\rangle \geq_0 \langle b, B, g\rangle$$

($\langle a, A, f\rangle$ is stronger than $\langle b, B, g\rangle$) iff

(1) $f \supseteq g$

(2) $a \supseteq b$

(3) $\pi_{\max(a),\max(b)}$"$A \subseteq B$.

We now define a forcing notion Q_n which is an extender analog of the one element Prikry forcing of 1.3.

Definition 2.8 $Q_n = Q_{n0} \cup Q_{n1}$.

Definition 2.9 The direct extension ordering $\leq^*$ on Q_n is defined to be $\leq_0 \cup \leq_1$.

Definition 2.10 Let $p, q \in Q_n$.

Then $p \leq q$ iff either

(1) $p \leq^* q$ or

(2) $p = \langle a, A, f\rangle \in Q_{n0}, q \in Q_{n1}$ and the following holds:

  (a) $q \supseteq f$

  (b) $\mathrm{dom}(q) \supseteq a$

  (c) $q(\max(a)) \in A$

  (d) for every $\beta \in a \quad q(\beta) = \pi_{\max(a),\beta}(q(\max(a)))$.

Clearly, the forcing $\langle Q_n, \leq\rangle$ is equivalent to $\langle Q_{n1}, \leq_1\rangle$, i.e. the Cohen forcing. However, the following basic facts relate it to the Prikry type forcing notion.

2.1. *Lecture 4*

Lemma 2.11 *$\langle Q_n, \leq^* \rangle$ is κ_n-closed.*

Lemma 2.12 *$\langle Q_n, \leq, \leq^* \rangle$ satisfies the Prikry condition, i.e. for every $p \in Q_n$ and every statement σ of the forcing language there is $q \geq^* p$ deciding σ.*

Proof. Let $p = \langle a, A, f \rangle$. Suppose otherwise. By recursion on $\nu \in A$ define an increasing sequence $\langle p_\nu \mid \nu \in A \rangle$ of elements of Q_{n1} with *dom* $p_\nu \cap a = \emptyset$ as follows. Suppose $\langle p_\rho \mid \rho \in A \cap \nu \rangle$ is defined and $\nu \in A$. Define p_ν as follows: Let $g = \bigcup_{\rho<\nu} p_\rho$. Then $g \in Q_{n1}$. Consider $q = \langle a, A, g \rangle$. Let $q^\frown\langle \nu \rangle = g \cup \{\langle \beta, \pi_{\max(a),\beta}(\nu) \rangle \mid \beta \in a\}$. If there is $p \geq_1 q^\frown\langle \nu \rangle$ deciding σ, then let p_ν be some such p restricted to $\lambda \backslash a$. Otherwise, set $p_\nu = g$. Note that there will always be a condition deciding σ.

Finally, let $g = \bigcup_{\nu \in A} p_\nu$.

Shrink A to a set $B \in U_{n\max(a)}$ so that $p_\nu^\frown\langle \nu \rangle = p_\nu \cup \{\langle \beta, \pi_{\max(a),\beta}(\nu) \rangle \mid \beta \in a\}$ decides σ the same way or does not decide σ at all, for every $\nu \in B$. By our assumption $\langle a, B, g \rangle \not\Vdash \sigma$. However, pick some $h \geq \langle a, B, g \rangle$, $h \in Q_{n1}$ deciding on σ. Let $h(\max(a)) = \nu$. Then, $p_\nu^\frown\langle \nu \rangle$ decides σ. But this holds then for every $\nu \in B$. Hence, already $\langle a, B, g \rangle$ decides σ. Contradiction.
□

Let us now define the main forcing of this section by putting the blocks of Q_n's together. This forcing is called the extender based Prikry forcing over a singular cardinal.

Definition 2.13 The set $\mathcal{P}$ consists of sequences $p = \langle p_n \mid n < \omega \rangle$ so that

(1) for every $n < \omega$, $p_n \in Q_n$,
(2) there is an $\ell(p) < \omega$ so that for every $n < \ell(p)$, $p_n \in Q_{n1}$, and for every $n \geq \ell(p)$, $p_n = \langle a_n, A_n, f_n \rangle \in Q_{n0}$ and $a_n \subseteq a_{n+1}$.

Definition 2.14 Let $p = \langle p_n \mid n < \omega \rangle$ and $q = \langle q_n \mid n < \omega \rangle \in \mathcal{P}$. We set $p \geq q$ ($p \geq^* q$) iff for every $n < \omega$, $p_n \geq_{Q_n} q_n$ ($p_n \geq^*_{Q_n} q_n$).

The forcing $\langle \mathcal{P}, \leq \rangle$ does not satisfy the κ^+-c.c. However:

Lemma 2.15 *$\langle \mathcal{P}, \leq \rangle$ satisfies the κ^{++}-c.c.*

Proof. Let $\{p(\alpha) \mid \alpha < \kappa^{++}\}$ be a set of elements of $\mathcal{P}$, with $p(\alpha) = \langle p(\alpha)_n \mid n < \omega \rangle$ and $p(\alpha)_n = \langle a(\alpha)_n, A(\alpha)_n, f(\alpha)_n \rangle$ for $n \geq \ell(p(\alpha))$. There is an $S \subseteq \kappa^{++}$ stationary such that for every $\alpha, \beta \in S$ the following holds:

(1) $\ell(p(\alpha)) = \ell(p(\beta)) = \ell$.
(2) for every $n < \ell$, $\{dom\ p(\alpha)_n \mid \alpha \in S\}$ forms a Δ-system with $p(\alpha)_n$ and $p(\beta)_n$ having the same values on its kernel.
(3) for every $n \geq \ell$, $\{(a(\alpha)_n \cup dom(f(\alpha)_n) \mid \alpha \in S\}$ forms a Δ-system with $f(\alpha)_n$, $f(\beta)_n$ having the same values on the kernel. Also, if $\alpha, \beta \in S$ then $a(\alpha)_n \cap dom(f(\beta)_n) = \emptyset$.

Now let $\alpha < \beta$ be in S. We construct a condition $q = \langle q_n \mid n < \omega \rangle$ stronger than both $p(\alpha)$ and $p(\beta)$.

For every $n < \ell$ let $q_n = p(\alpha)_n \cup p(\beta)_n$. Now suppose that $n \geq \ell$. q_n will be of the form $\langle b_n, B_n, g_n \rangle$. Set $g_n = f(\alpha)_n \cup f(\beta)_n$. We would like to define b_n as the union of $a(\alpha)_n$ and $a(\beta)_n$. But 2.6(2(iii)) requires the existence of a maximal element in the $\leq_{E_n}$ order which need not be the case in the simple union of $a(\alpha)_n$ and $a(\beta)_n$. It is easy to fix this. Just pick some $\rho < \lambda$ above $a(\alpha)_n \cup a(\beta)_n$ in the $\leq_{E_n}$ order. Also let $\rho > \sup(dom(f(\alpha)_n)) + \sup(dom(f(\beta))_n)$. Lemma 2.2 insures that there are such ρ's. Now we set $b_n = a(\alpha)_n \cup a(\beta)_n \cup \{\rho\}$. Let $B'_n = \pi^{-1}_{\rho\alpha^*}(A(\alpha)_n) \cap \pi^{-1}_{\rho\beta^*}(A(\beta)_n)$, where $\alpha^* = \max(a(\alpha)_n)$ and $\beta^* = \max(a(\beta)_n)$. Finally we shrink B'_n to a a set $B_n \in U_{n\rho}$ satisfying 2.6((4), (5)). This is possible by Lemmas 2.3 and 2.4. $\square$

For $p = \langle p_n \mid n < \omega \rangle \in \mathcal{P}$ set $p \upharpoonright n = \langle p_m \mid m < n \rangle$ and $p \backslash n = \langle p_m \mid m \geq n \rangle$. Let $\mathcal{P} \upharpoonright n = \{p \upharpoonright n \mid p \in \mathcal{P}\}$ and $\mathcal{P} \backslash n = \{p \backslash n \mid p \in \mathcal{P}\}$. Then the following lemmas are obvious:

Lemma 2.16 *$\mathcal{P} \simeq \mathcal{P} \upharpoonright n \times \mathcal{P} \backslash n$ for every $n < \omega$.*

Lemma 2.17 *$\langle \mathcal{P} \backslash n, \leq^* \rangle$ is κ_n-closed. Moreover if $\langle p^\alpha \mid \alpha < \delta < \kappa \rangle$ is a $\leq^*$ increasing sequence with $\kappa_{\ell(p_0)} > \delta$, then there is $p \geq^* p^\alpha$ for every $\alpha < \delta$.*

We will turn now to the Prikry condition and establish a more general statement which will allow us to deduce in addition that κ^+ is preserved after forcing with $\langle \mathcal{P}, \leq \rangle$.

Let us introduce first some notation. For $p = \langle p_n \mid n < \omega \rangle \in \mathcal{P}$ and m with $\ell(p) \leq m < \omega$, let $p_m = \langle a_m, A_m, f_m \rangle$. Denote a_m by $a_m(p)$, A_m by $A_m(p)$ and f_m by $f_m(p)$. Let $\langle \nu_{\ell(p)}, \ldots, \nu_m \rangle \in \prod_{k=\ell(p)}^m A_k(p)$. Denote by $p^\frown \langle \nu_{\ell(p)}, \ldots, \nu_m \rangle$ the condition obtained from p by adding the sequence $\langle \nu_{\ell(p)}, \ldots, \nu_m \rangle$, i.e. a condition $q = \langle q_n \mid n < \omega \rangle$ such that $q_n = p_n$ for every n, $n < \ell(p)$ or $n > m$, and if $\ell(p) \leq n \leq m$ then $q_n = f_n(p) \cup \{\langle \beta, \pi_{\max(a_n(p))\beta}(\nu_n) \rangle \mid \beta \in a_n(p)\}$.

We prove the following analog of 1.13:

Lemma 2.18 *Let $p \in \mathcal{P}$ and D be a dense open subset of $\langle \mathcal{P}, \leq \rangle$ above p. Then there are $p^* \geq^* p$ and $n^* < \omega$ such that for every $\langle \nu_0, \ldots, \nu_{n^*-1} \rangle \in \prod_{m=\ell(p)}^{\ell(p)+n^*-1} A_m(p^*)$,*

$$p^{*\frown} \langle \nu_0, \ldots, \nu_{n^*-1} \rangle \in D.$$

Let us first deduce the Prikry condition from this lemma.

Lemma 2.19 *Let $p \in \mathcal{P}$ and σ be a statement of the forcing language. Then there is a $p^* \geq^* p$ deciding σ.*

Proof of 2.19 from 2.18 Consider $D = \{q \in \mathcal{P} \mid q \geq p \text{ and } q \parallel \sigma\}$. Clearly, D is dense open above p. Apply 2.18 to this D and choose n^* as small as possible and $p^* \geq^* p$ such that for every $q \geq p^*$ with $\ell(q) \geq n^*$, $q \in D$. If $n^* = \ell(p)$, then we are done. Suppose otherwise. Assume for simplicity that $\ell(p) = 0$ and $n^* = 2$. Then let $p^* = \langle p_n^* \mid n < \omega \rangle$ and for every $n < \omega$ let $p_n^* = \langle a_n^*, A_n^*, f_n^* \rangle$. Let $\alpha_0 = \max(a_0^*)$ and $\alpha_1 = \max(a_1^*)$. Then $A_0^* \in U_{0\alpha_0}$ and $A_1^* \in U_{1\alpha_1}$. Let $\nu_0 \in A_0^*$ and $\nu_1 \in A_1^*$. Consider $p^{*\frown}\langle \nu_0, \nu_1 \rangle$ the condition obtained from p^* by adding ν_0 and ν_1. Clearly, $\ell(p^{*\frown}\langle \nu_0, \nu_1 \rangle) = 2$. Hence it decides σ. Now we shrink A_1^* to $A_{1\nu_0}^*$ so that for every $\nu_1', \nu_1'' \in A_{1\nu_0}^*$ $p^{*\frown}\langle \nu_0, \nu_1' \rangle$ and $p^{*\frown}\langle \nu_0, \nu_1'' \rangle$ decide σ the same way. Let $A_1^{**} = \bigcap \{A_{1\nu_0}^* \mid \nu_0 \in A_0^*\}$. We shrink now A_0^* to A_0^{**} so that for every $\nu_0', \nu_0'' \in A_0^{**}$ and for every $\nu_1 \in A_1^{**}$ $p^{*\frown}\langle \nu_0', \nu_1 \rangle$ and $p^{*\frown}\langle \nu_0'', \nu_1 \rangle$ decide σ in the same way. Let p^{**} be a condition obtained from p^* by replacing in it A_0^* by A_0^{**} and A_1^* by A_1^{**}. Then $p^{**} \geq^* p^*$ and $p^{**} \| \sigma$. Contradiction.

2.2. *Lecture 5 — Proof of 2.18*

The main objective is to reduce the problem to the point where we can use the argument of the corresponding fact in Sec. 1.3, as if we were forcing using $\langle U_{n\max(a_n)} \mid n < \omega \rangle$.

We first prove the following crucial claim:

Claim 1 (2.18.1) There is $p' \geq^* p$, $p' = \langle p_n' \mid n < \omega \rangle$ such that for every $q \geq p'$ with $q = \langle q_n \mid n < \omega \rangle$, if $q \in D$ then also

$$\langle p_n' \mid n < \ell(p) \rangle^\frown \langle q_n \restriction a_n(p') \cup f_n(p') \mid \ell(p') \leq n < \ell(q) \rangle^\frown \langle p_n' \mid n \geq \ell(q) \rangle \in D\,,$$

where $p_n' = \langle a_n(p'), A_n(p'), f_n(p') \rangle$ for $n \geq \ell(p)$.

Proof. Choose a function $h : \kappa \leftrightarrow [\kappa]^{<\omega}$, such that for every $n < \omega$, $h \restriction \kappa_n : \kappa_n \leftrightarrow [\kappa_n]^{<\omega}$. Now define by recursion a $\leq^*$-increasing sequence $\langle p^\alpha \mid \alpha < \kappa \rangle$ of direct extensions of p, where $p^\alpha = \langle p_n^\alpha \mid n < \omega \rangle$ and, for $n \geq \ell(p)$, $p_n^\alpha = \langle a_n^\alpha, A_n^\alpha, f_n^\alpha \rangle$. Set $p^0 = p$. Suppose that $\alpha < \kappa$ and $\langle p^\beta \mid \beta < \alpha \rangle$ has been defined. As a recursive assumption we assume the following:

$$(*) \qquad \text{for every } n < \omega \text{ and for } \beta, \gamma, \kappa_n \leq \beta, \gamma < \kappa, \text{ if } \ell(p) \leq m \leq n+1$$

$$\text{then } a_m^\beta = a_m^\gamma \text{ and } A_m^\beta = A_m^\gamma .$$

Let $\tilde{p}^\alpha$ be $p^{\alpha-1}$ if α is successor ordinal, and a direct extension of $\langle p^\beta \mid \beta < \alpha \rangle$ satisfying (*) if α is a limit ordinal. Note that if $n < \omega$ is the maximal such that $\alpha \geq \kappa_n$ then 2.17 applies, since the parts of p^β's below κ_{n+1} satisfy (*). Now we consider $h(\alpha)$. Let $h(\alpha) = \langle \nu_1, \ldots, \nu_k \rangle$. If $\langle \nu_0, \ldots, \nu_{k-1} \rangle \notin \prod_{m=\ell(p)}^{\ell(p)+k-1} A_m(\tilde{p}^\alpha)$ then we set $p^\alpha = \tilde{p}^\alpha$, where for $m \geq \ell(p)$, $\tilde{p}_m^\alpha = \langle a_m(\tilde{p}^\alpha), A_m(\tilde{p}^\alpha), f_m(\tilde{p}^\alpha) \rangle$. Otherwise we consider $q = \tilde{p}^{\alpha\frown} \langle \nu_0, \ldots, \nu_{k-1} \rangle$. If there is no direct extension of q inside D, then let $p^\alpha = \tilde{p}^\alpha$. Otherwise let $s = \langle s_n \mid n < \omega \rangle \geq^* q$ be in D. Define $p^\alpha = \langle p_n^\alpha \mid n < \omega \rangle$ then as follows:

for each $n \geq \ell(p) + k$ or $n < \ell(p)$ set $p_n^\alpha = s_n$;
for each n with $\ell(p) \leq n \leq \ell(p) + k - 1$ let $a_n(p^\alpha) = a_n(\tilde{p}^\alpha)$, $A_n(p^\alpha) = A_n(\tilde{p}^\alpha)$ and $f_n(p^\alpha) = f_n(s) \restriction ((dom\ f_n(s)) \backslash a_n(\tilde{p}^\alpha))$.

The meaning of this last part of the definition is that we extend for n with $\ell(p) \leq n \leq \ell(p) + k - 1$ only $f_n(\tilde{p}^\alpha)$ and only outside of $a_n(\tilde{p}^\alpha)$. Clearly such defined p^α satisfies (*).

Finally (*) allows us to put all the $\langle p^\alpha \mid \alpha < \kappa \rangle$ together. Thus we define $p' = \langle p'_n \mid n < \omega \rangle$:

$$\text{if} \quad n < \ell(p) \quad \text{then} \quad p'_n = \bigcup_{\alpha < \kappa} p_n^\alpha$$

$$\text{if} \quad n \geq \ell(p) \quad \text{then} \quad f_n(p') = \bigcup_{\alpha < \kappa} f_n(p^\alpha)$$

$$a_n(p') = a_n(p^{\kappa_n}) \quad \text{and} \quad A_n(p') = A_n(p^{\kappa_n}) .$$

Obviously $p' \in \mathcal{P}$ and $p' \geq^* p$. This p' is as desired. Thus, if $q \geq p'$ is in D, then we consider $\alpha = h^{-1}(\langle q_n(\max(a_n(p'))) \mid \ell(p) \leq n < \ell(q) \rangle)$. By the construction of $p^\alpha \leq^* p'$, $p^{\alpha\frown} \langle q_n(\max(a_n(p'))) \mid \ell(p) \leq n < \ell(q) \rangle$ will be in D. Then also $p'^\frown \langle q_n(\max(a_n(p')) \mid \ell(p) \leq n < \ell(q) \rangle \in D$, since D is open. $\square$ of the claim.

Now let $p' \geq^* p$ be given by the claim. Assume for simplicity that $\ell(p) = 0$. We would like to shrink the sets $A_n(p')$ in a certain way. Thus define $p(1) \geq^* p'$ such that

$(*)_1$ for every $m < \omega$, $\langle \nu_0, \ldots, \nu_{m-1} \rangle \in \prod_{n=0}^{m-1} A_n(p(1))$, if for some $\nu \in A_m(p(1))$ $p(1)^\frown \langle \nu_0, \ldots, \nu_{m-1}, \nu \rangle \in D$ then for every $\nu' \in A_m(p(1))$ $p(1)^\frown \langle \nu_0, \ldots, \nu_{m-1}, \nu' \rangle \in D$.

Let $m < \omega$ and $\vec{\nu} = \langle \nu_0, \ldots, \nu_{m-1} \rangle \in \prod_{n=0}^{m-1} A_n(p')$, where in case of $m = 0$, $\vec{\nu}$ is the empty sequence. Consider the set $X_{m,\vec{\nu}} = \{\nu \in A_m(p') \mid p'^\frown \langle \nu_0, \ldots, \nu_{m-1}, \nu \rangle \in D\}$. Define $A_{m\vec{\nu}}$ to be $X_{m,\vec{\nu}}$, if $X_{m\vec{\nu}} \in U_{m,\max(a_m(p'))}$ and $A_m(p') \backslash X_{m,\vec{\nu}}$, otherwise. Let $A_m = \bigcap \{A_{m,\vec{\nu}} \mid \vec{\nu} \in \prod_{n=0}^{m-1} A_n(p')\}$. Define now $p(1) = \langle p(1)_n \mid n < \omega \rangle$ as follows: for each $n < \omega$ let $p(1)_n = \langle a_n(p'), A_n, f(p') \rangle$. Clearly, such defined $p(1)$ satisfies $(*)_1$.

Then, in the similar fashion we chose $p(2) \geq^* p(1)$ satisfying

$(*)_2$ for every $m < \omega$, $\langle \nu_0, \ldots, \nu_{m-1} \rangle \in \prod_{n=0}^{m-1} A_n(p(2))$, if for some $\langle \nu_m, \nu_{m+1} \rangle \in A_m(p(2)) \times A_{m+1}(p(2))$, $p(2)^\frown \langle \nu_0, \ldots, \nu_{m-1} \rangle^\frown \langle \nu_m, \nu_{m+1} \rangle \in D$ then for every $\langle \nu'_m, \nu'_{m+1} \rangle \in A_m(p(2)) \times A_{m+1}(p(2))$

$$p(2)^\frown \langle \nu_0, \ldots, \nu_{m-1} \rangle^\frown \langle \nu'_m, \nu'_{m+1} \rangle \in D .$$

Continue and define for every k, $2 \leq k < \omega$, $p(k) \geq^* p(k-1)$ satisfying $(*)_k$, where $(*)_k$ is defied analogously for k-sequences. Finally let p^* be a direct extension of $\langle p(k) \mid 1 \leq k < \omega \rangle$. Let $s \geq p^*$ be in D. Set $n^* = \ell(s)$. Consider $\langle s_0(\max(a_0(p^*))), \ldots, s_{n^*-1}(\max(a_{n^*-1}(p^*))) \rangle$. Then the choice of p', $p' \leq^* p^*$ and openness of D imply that $p^{*\frown} \langle s_0(\max(a_0(p^*)), \ldots, s_{n^*-1}(\max(a_{n^*-1}(p^*))) \rangle$ is in D. But $p^* \geq^* p(n^*)$. So, p^* satisfies $(*)_{n^*}$. Hence, for every $\langle \nu_0, \ldots, \nu_{n^*-1} \rangle \in \prod_{m=0}^{n^*-1} A_m(p^*)$ $p^{*\frown} \langle \nu_0, \ldots, \nu_{n^*-1} \rangle \in D$.
□

Combining these lemmas we obtain the following:

Proposition 2.20 *The forcing $\langle \mathcal{P}, \leq \rangle$ does not add new bounded subsets to κ and preserves all the cardinals above κ^+.*

Actually, it is not hard now to show that κ^+ is preserved as well.

Lemma 2.21 *Forcing with $\langle \mathcal{P}, \leq \rangle$ preserves κ^+.*

Proof. Suppose that $(\kappa^+)^V$ is not a cardinal in a generic extension $V[G]$. Recall that $cf(\kappa) = \aleph_0$ and by 2.20 it is preserved. So, $cf((\kappa^+)^V) < \kappa$ in

$V[G]$. Pick $p \in G$, $\delta < \kappa$ and a name $\underset{\sim}{g}$ so that $\kappa_{\ell(p)} > \delta$ and

$$p \Vdash \left(\underset{\sim}{g} : \check{\delta} \to (\kappa^+)^V \quad \text{and} \quad \operatorname{rng} \underset{\sim}{g} \quad \text{is unbounded in} \quad (\kappa^+)^V\right) .$$

For every $\tau < \delta$ let $D_\tau = \{q \in \mathcal{P} \mid q \geq p$ and for some $\alpha < \kappa^+$ $q \Vdash \underset{\sim}{g}(\check{\tau}) = \check{\alpha}\}$. Define by recursion, using 2.17, a $\leq^*$-increasing sequence $\langle p^\tau \mid \tau < \delta \rangle$ of $\leq^*$-extensions of p so that p^τ satisfies the conclusion 2.18 with $D = D_\tau$. By 2.17, there is $p^\delta \geq^* p^\tau$ for each $\tau < \delta$.

Now let $\tau < \delta$. By the choice of p^τ there is $n(\tau) < \omega$ such that for every

$$\langle \nu_0, \dots, \nu_{n(\tau)-1} \rangle \in \prod_{m=\ell(p)}^{\ell(p)+n(\tau)-1} A_m(p^\delta)$$

$$p^{\delta \frown} \langle \nu_0, \dots, \nu_{n(\tau)-1} \rangle \in D_\tau .$$

This means that for some $\alpha(\nu_0, \dots, \nu_{n(\tau)-1}) < \kappa^+$

$$p^{\delta \frown} \langle \nu_0, \dots, \nu_{n(\tau)-1} \rangle \Vdash \underset{\sim}{g}(\check{\tau}) = \check{\alpha}(\nu_0, \dots, \nu_{n(\tau)-1}) .$$

Set

$$\alpha(\tau) = \sup\{\alpha(\nu_0, \dots, \nu_{n(\tau)-1}) \mid \langle \nu_0, \dots, \nu_{n(\tau)-1} \rangle \in \textstyle\prod_{m=\ell(p)}^{\ell(p)+n(\tau)-1} A_m(p^\delta)\}.$$

Then, clearly, $\alpha(\tau) < \kappa^+$ and

$$p^\delta \Vdash \underset{\sim}{g}(\check{\tau}) < \check{\alpha}(\tau).$$

Now let $\alpha^* = \bigcup_{\tau<\delta} \alpha(\tau)$. Then again $\alpha^* < \kappa^+$ and

$$p^\delta \Vdash \forall \tau < \check{\delta} \quad (\underset{\sim}{g}(\tau) < \check{\alpha}^*).$$

But this is impossible since $p \leq^* p^\delta$ forced that the range of g was unbounded in κ^+. Contradiction.

□

Finally, let us show that this forcing adds λ ω-sequences to κ. Thus, let $G \subseteq \mathcal{P}$ be generic. For every $n < \omega$ define a function $F_n : \lambda \to \kappa_n$ as follows:

$F_n(\alpha) = \nu$ if for some $p = \langle p_m \mid m < \omega \rangle \in G$ with $\ell(p) > n$, $p_n(\alpha) = \nu$.

Now for every $\alpha < \lambda$ set $t_\alpha = \langle F_n(\alpha) \mid n < \omega \rangle$. Let us show that the set $\{t_\alpha \mid \alpha < \lambda\}$ has cardinality λ. Notice that we cannot claim that all such sequences are new or even distinct due to Cohen parts of conditions, i.e. f_n's.

Lemma 2.22 *For every $\beta < \lambda$ there is $\alpha, \beta < \alpha < \lambda$ such that t_α dominates every t_γ with $\gamma \leq \beta$.*

Proof. Suppose otherwise. Then there is $p = \langle p_n \mid n < \omega \rangle \in G$ and $\beta < \lambda$ such that

$$p \Vdash \forall \alpha (\beta < \alpha < \lambda \rightarrow \exists \gamma \leq \beta \ (\underset{\sim}{t_\alpha} \quad \text{does not dominate} \quad \underset{\sim}{t_\gamma})).$$

For every $n \geq \ell(p)$ let $p_n = \langle a_n, A_n, f_n \rangle$. Pick some $\alpha \in \lambda \backslash \Big(\bigcup_{n<\omega} a_n \cup \bigcup \mathrm{dom}(f_n) \cup (\beta + 1) \Big)$. We extend p to a condition q so that $q \geq^* p$ and for every $n \geq \ell(q) = \ell(p)$, $\alpha \in b_n$, where $q_n = \langle b_n, B_n, g_n \rangle$. Then q will force that t_α dominates every t_γ with $\gamma < \alpha$. This leads to the contradiction. Thus, let $\gamma < \alpha$ and assume that q belongs to the generic subset of $\mathcal{P}$. Then either $t_\gamma \in V$ or it is a new ω-sequence. If $t_\gamma \in V$ then it is dominated by t_α by the usual density arguments. If t_γ is new, then for some $r \geq q$ in the generic set $\gamma \in c_n$ for every $n \geq \ell(r)$, where $r_n = \langle c_n, C_n, h_n \rangle$. But also $\alpha \in c_n$ since $c_n \supseteq b_n$. This implies $F_n(\alpha) > F_n(\gamma)$ (by 2.6(5)) and we are done.

□

Now we have the following conclusion.

Theorem 2.23 *The following holds in $V[G]$:*

(a) all cardinals and cofinalities are preserved.

(b) no new bounded subsets are added to κ; in particular GCH holds below κ.

(c) there are λ new ω sequences in $\prod_{n<\omega} \kappa_n$. In particular $2^\kappa \geq \lambda$.

Remark 2.24 The initial large cardinal assumptions used here are not optimal.

It is tempting to extend 2.22 and claim that $\langle t_\alpha \mid \alpha < \lambda$ and $t_\alpha \notin V \rangle$ is a scale in $\prod_{n<\omega} \kappa_n$, i.e. for every $t \in \prod_{n<\omega} \kappa_n$ there is $\alpha < \lambda$ such $t_\alpha \notin V$ and t_α dominates t. Unfortunately this is not true in general. We need to replace $\prod_{n<\omega} \kappa_n$ by the product of a sequence $\langle \lambda_n \mid n < \omega \rangle$ related to λ (basically the Prikry sequence for $U_{n\lambda}$ whenever it is defined). Assaf Sharon made a full analysis of possible cofinalities structure for a similar forcing (the one that will be discussed in the next section). Let us now deal with a special case that cannot be covered by such forcing. Let us assume that for every $n < \omega$, $j_n(\kappa_n) = \lambda$, where $j_n : V \to M_n \simeq Ult(V, E_n)$ is the canonical embedding. In particular each κ_n is a superstrong cardinal. Then the following holds.

Lemma 2.25 *Let $t \in \prod_{n<\omega} \kappa_n$ in $V[G]$. Then there is $\alpha < \lambda$ such that $t_\alpha \notin V$ and for all but finitely many $n < \omega$, $t_\alpha(n) > t(n)$.*

Proof. Let $\underset{\sim}{t}$ be a name of t. Pick $p \in G$ forcing "$\underset{\sim}{t} \in \prod_{n<\omega} \check{\kappa}_n$". Define for every $n < \omega$ a set dense open above p:

$$D_n = \{q \in \mathcal{P} \mid q \geq p \quad \text{and there is} \quad \nu_n < \kappa_n \quad \text{such that} \quad q \Vdash \underset{\sim}{t}(n) = \check{\nu}_n\}\,.$$

Apply 2.18 to each of D_n's and construct a $\leq^*$-sequence $\langle p(k) \mid k < \omega\rangle$ of direct extensions of p such that $p(k)$ and D_k satisfy the conclusion of 2.18. Let p^* be a common direct extension of $p(k)$'s. Then for every $k, 1 \leq k < \omega$, there is $n(k) < \omega$ such that for every $\langle \nu_0, \ldots, \nu_{n(k)-1}\rangle \in \prod_{m=\ell(p)}^{\ell(p)+n(k)-1} A_m(p^*)$

$$p^{*\frown}\langle \nu_0, \ldots, \nu_{n(k)-1}\rangle \Vdash \underset{\sim}{t}(k-1) = \check{\xi}(\nu_0, \ldots, \nu_{n(k)-1})$$

for some $\xi(\nu_0, \ldots, \nu_{n(k)-1}) < \kappa_{k-1}$. Assume for simplicity of notation that $\ell(p) = 0$. Let $1 \leq k < \omega$. We can assume that $\xi(\nu_0, \ldots, \nu_{n(k)-1})$, defined above, depends really only on $\nu_0, \ldots, \nu_{k-1}$, since its values are below κ_{k-1} and ultrafilters over κ_m's are κ_k-complete for $m \geq k$. Also assume that for every $m > 0$ $A_m(p^*) \cap \kappa_{m-1} = \emptyset$. Now, we replace ξ by a bigger function η depending only on ν_{k-1}. Thus set

$$\eta(\nu_{k-1}) = \cup\left\{\xi(\nu_0, \ldots, \nu_{k-2}, \nu_{k-1}) \mid \langle \nu_0, \ldots, \nu_{k-2}\rangle \in \prod_{m=0}^{k-2} A_m(p^*)\right\} + \nu_{k-1}\,.$$

Clearly, $\eta(\nu_{k-1}) < \kappa_{k-1}$. So,

$$p^{*\frown}\langle \nu_0, \ldots, \nu_{k-1}\rangle \Vdash \underset{\sim}{t}(k-1) < \check{\eta}(\nu_{k-1})$$

for every k, $1 \leq k < \omega$ and every $\langle \nu_0, \ldots, \nu_{k-1}\rangle \in \prod_{m=0}^{k-1} A_m(p^*)$. For every $n < \omega$ let $\eta_n : A_n(p^*) \to \kappa_n$ be the restriction of η to κ_n. Let $\alpha_n = \max(a_n(p^*))$. Consider $j_n(\eta_n)(\alpha_n)$ where $j_n : V \to M_n$ is the embedding of the extender E_n. Then $j_n(\eta_n)(\alpha_n) < j_n(\kappa_n) = \lambda$. Choose some α below λ

and above $\bigcup_{n<\omega} j_n(\eta_n)(\alpha_n) \cup (dom\ f_n(p^*))$. Now extend p^* to a condition p^{**} such that $p^{**} \geq^* p^*$ and for every $n < \omega$ $\alpha \in a_n(p^{**})$. Then,

$$p^{**} \Vdash \forall n(\underset{\sim}{t}_\alpha(n) > \eta_n(\underset{\sim}{t}_{\alpha_n}(n)) > \underset{\sim}{t}(n)).$$

So we are done.

□

The extender based forcing described in this section can also be used with much stronger extenders than those used here. Thus with minor changes we can deal with E_n's such that $j_n(\kappa_n) < \lambda$ but requiring $j_n(\kappa_{n+1}) > \lambda$. Once $j_n(\kappa_{n+1}) \leq \lambda$ for infinitely many n's then the arguments like one in the proof of the Prikry condition seem to break down completely.

3. Lecture 6 — A Short Extender Forcing Gap 2

We assume GCH. Let $\kappa = \bigcup_{n<\omega} \kappa_n$, $\kappa_0 < \kappa_1 < \cdots < \kappa_n < \cdots$ and for every $n < \omega$ κ_n is $\lambda_n + 1$-strong, where $\lambda_n = \kappa_n^{+n+2}$.

We would like to blow up the power of κ to κ^{++} without adding new bounded subsets to κ.

For each $n < \omega$ we fix an extender E_n witnessing $\lambda_n + 1$-strongness of κ_n. Let $j_n : V \to M \simeq Ult(V, E_n)$. We define ultrafilters $U_{n\alpha}(\alpha < \lambda_n)$ as before by setting $X \in U_{n\alpha}$ iff $\alpha \in j_n(X)$. Also the order $\leq_{E_n}$ over λ_n is defined as in Sec. 2. Here λ will be just κ^{++}. The first idea for blowing power of κ to λ is to simulate the forcing $\mathcal{P}$ of Sec. 2. It was built from blocks Q_n's. The essential part of Q_n is Q_{n0} which typical element has a form $\langle a, A, f\rangle$, where f is a Cohen condition, A is a set of measure one, but the main and problematic part $a \subset \lambda$ is actually a set of indexes of the extender E_n. E_n had length λ in Sec. 2 but now it is very short. Its length is $\lambda_n < \kappa_{n+1} < \kappa$. Here we take a to be an order preserving function from λ into the set of indexes of E_n, i.e. into λ_n. Formally:

Definition 3.1 Let Q_{n0} be the set of triples $\langle a, A, f\rangle$ so that

(1) $f \in Q_{n1}$, where Q_{n1} is defined as in 2.5.

(2) a is a partial order preserving function from κ^{++} to λ_n such that

  (i) $|a| < \kappa_n$

  (ii) $\mathrm{dom}(a) \cap \mathrm{dom}(f) = \emptyset$

  (iii) $\mathrm{rng}(a)$ has a $\leq_{E_n}$-maximal element

(3) $A \in U_{n\,\max(\mathrm{rng}(a))}$

(4) for every $\alpha, \beta, \gamma \in \mathrm{rng}(a)$, if $\alpha \geq_{E_n} \beta \geq_{E_n} \gamma$ then

$$\pi_{\alpha\gamma}(\rho) = \pi_{\beta\gamma}(\pi_{\alpha\beta}(\rho))$$

for every $\rho \in \pi''_{\max(\mathrm{rng}(a)),\alpha} A$

(5) for every $\alpha > \beta$ in $\mathrm{rng}(a)$ and $\nu \in A$

$$\pi_{\max(\mathrm{rng}(a)),\alpha}(\nu) > \pi_{\max(\mathrm{rng}(a)),\beta}(\nu) .$$

Definition 3.2 Let $\langle a, A, f\rangle, \langle b, B, g\rangle \in Q_{n0}$. Then

$$\langle a, A, f\rangle \geq_0 \langle b, B, g\rangle$$

($\langle a, A, f\rangle$ is stronger than $\langle b, B, g\rangle$) iff

(1) $f \supseteq g$

(2) $a \supseteq b$

(3) $\pi_{\max(\mathrm{rng}(a)),\max(\mathrm{rng}(b))}$ "$A \subseteq B$.

Definition 3.3 $Q_n = Q_{n0} \cup Q_{n1}$.

Definition 3.4 The direct extension ordering $\leq^*$ on Q_n is defined to be $\leq_0 \cup \leq_1$.

Definition 3.5 Let $p, q \in Q_n$.
Then $p \leq q$ iff either

(1) $p \leq^* q$ or

(2) $p = \langle a, A, f\rangle \in Q_{n0}, q \in Q_{n1}$ and the following holds:
  (a) $q \supseteq f$
  (b) $\mathrm{dom}(q) \supseteq \mathrm{dom}(a)$
  (c) $q(\max(\mathrm{dom}(a))) \in A$
  (d) for every $\beta \in \mathrm{dom}(a)$ $q(\beta) = \pi_{\max(\mathrm{rng}(a)),a(\beta)}(q(\max(a)))$.

Lemmas 2.11 and 2.12 are valid with proofs requiring minor changes. The forcing $\mathcal{P}$ of 2.13 is defined here similarly:

Definition 3.6 The set $\mathcal{P}$ consists of sequences $p = \langle p_n \mid n < \omega\rangle$ so that

(1) for every $n < \omega$ $p_n \in Q_n$

(2) there is $\ell(p) < \omega$ so that for every $n < \omega$ $p_n \in Q_{n1}$, for every $n \geq \ell(p)$ $p_n = \langle a_n, A_n, f_n\rangle$ and $\mathrm{dom}(a_n) \subseteq \mathrm{dom}(a_{n+1})$.

Definition 3.7 Let $p = \langle p_n \mid n < \omega \rangle$, $q = \langle q_n \mid n < \omega \rangle \in \mathcal{P}$. We define $p \geq q$ $(p \geq^* q)$ iff for every $n < \omega$ $p_n \geq_{Q_n} q_n$ $(p_n \geq^*_{Q_n} q_n)$.

For $p = \langle p_n \mid n < \omega \rangle \in \mathcal{P}$ let $p \restriction n = \langle p_m \mid m < n \rangle$ and $p \backslash n = \langle p_m \mid m \geq n \rangle$. Set $\mathcal{P} \restriction n = \{p \restriction n \mid p \in \mathcal{P}\}$ and $\mathcal{P} \backslash n = \{p \backslash n \mid p \in \mathcal{P}\}$.

The following lemmas are obvious:

Lemma 3.8 *$\mathcal{P} \simeq \mathcal{P} \restriction n \times \mathcal{P} \backslash n$ for every $n < \omega$.*

Lemma 3.9 *$\langle \mathcal{P} \backslash n, \leq^* \rangle$ is κ_n-closed.*

The proof of the Prikry condition is the same as 2.18.

Lemma 3.10 *$\langle \mathcal{P}, \leq, \leq^* \rangle$ satisfies the Prikry condition.*

The ω-sequences $t_\alpha = \langle F_\alpha(n) \mid n < \omega \rangle$ defined as in Sec. 1 will witness that λ new ω-sequences are added by $\langle \mathcal{P}, \leq \rangle$. Thus we obtain the following:

Proposition 3.11 *The forcing $\langle \mathcal{P}, \leq \rangle$ does not add new bounded subsets to κ and it adds $\lambda = (\kappa^{++})^V$ new ω-sequences to κ.*

The problem is that κ^{++}-c.c. fails badly. Thus, any two conditions p and q such that for infinitely many n's $\mathrm{rng}(a_n(p)) = \mathrm{rng}(a_n(q))$ but $\mathrm{dom}(a_n(p)) \neq \mathrm{dom}(a_n(q))$ are incompatible. Using this it is possible to show that $\langle \mathcal{P}, \leq \rangle$ collapses λ to κ^+. The rest of the lecture will be devoted to the task of repairing the chain condition. Thus we shall identify various conditions in $\mathcal{P}$.

Fix $n < \omega$. For every $k \leq n$ we consider a language $\mathcal{L}_{n,k}$ containing two relation symbols, a function symbol, a constant c_α for every $\alpha < \kappa_n^{+k}$ and constants c_{λ_n}, c. Consider a structure

$$\mathfrak{a}_{n,k} = \langle H(\chi^{+k}), \in, E_n, \quad \text{the enumeration of} \quad [\lambda_n]^{<\lambda_n} \text{ (Sec. 2 (p. 17))},$$

$$\lambda_n, \chi, 0, 1, \ldots, \alpha \ldots \mid \alpha < \kappa_n^{+k} \rangle$$

in this language, where χ is a regular cardinal large enough. For an ordinal $\xi < \chi$ (usually ξ will be below λ_n) we denote by $tp_{n,k}(\xi)$ the $\mathcal{L}_{n,k}$-type realized by ξ in $\mathfrak{a}_{n,k}$.

Let $\mathcal{L}'_{n,k}$ be the language obtained from $\mathcal{L}_{n,k}$ by adding a new constant c'. For $\delta < \chi$ let $\mathfrak{a}_{n,k,\delta}$ be the $\mathcal{L}'_{n,k}$-structure obtained from $\mathfrak{a}_{n,k}$ by interpreting c' as δ. The type $tp_{n,k}(\delta, \xi)$ is the $\mathcal{L}'_{n,k}$-type realized by ξ in

$\mathfrak{a}_{n,k,\delta}$. Further, we shall identify types with ordinals corresponding to them in some fixed well-ordering of the power sets of κ_n^{+k}'s.

Definition 3.12 Let $k \leq n$ and $\beta < \lambda_n$. β is called k-good iff

(1) for every $\gamma < \beta$ $tp_{n,k}(\gamma, \beta)$ is realized unboundedly many times below λ_n;
(2) for every $a \subseteq \beta$ if $|a| < \kappa_n$ then there is $\alpha < \beta$ corresponding to a in the enumeration of $[\lambda_n]^{<\lambda_n}$.

β is called good if it is k-good for some $k \leq n$.

Further we will be interested mainly in k-good ordinals for $k > 2$. If $\alpha, \beta < \lambda_n = \kappa_n^{+n+2}$ realize the same k-type for $k > 2$, then $U_{n\alpha} = U_{n\beta}$, since the number of different $U_{n\alpha}$'s is κ_n^{++}.

Lemma 3.13 *The set* $\{\beta < \lambda_n \mid \beta \text{ is } n\text{-good}\} \cup \{\beta < \lambda_n \mid cf\beta < \kappa_n\}$ *contains a club.*

Proof. Let us show first that the set $\{\beta < \lambda_n \mid \forall \gamma < \beta \ \ tp_{n,n}(\gamma, \beta)$ is realized unboundedly often$\}$ contains a club. Suppose otherwise. Let S be a stationary set of β's such that there is $\gamma_\beta < \beta$ with $tp(\gamma_\beta, \beta)$ realized only boundedly many times below λ_n. Shrink S to a stationary S^* on which all γ_β's are the same. Let $\gamma_\beta = \gamma$ for every $\beta \in S^*$. The total number of n-types over γ, i.e. $tp_{n,n}(\gamma, -)$ is $\kappa_n^{+n+1} < \lambda_n$. Hence, there is a stationary $S^{**} \subseteq S^*$ such that for every $\alpha, \beta \in S^{**}$ $tp_{n,n}(\gamma, \alpha) = tp_{n,n}(\gamma, \beta)$. In particular the type $tp_{n,n}(\gamma, \beta)$ is realized unboundedly often below λ_n.

Contradiction.

Now, in order to finish the proof, notice that whenever $N \prec \mathfrak{a}_{n,n}$, $\beta = N \cap \lambda_n < \lambda_n$ and ${}^{\kappa_n >}N \subseteq N$ then β satisfies (2) of 3.12.
□

Lemma 3.14 *Suppose that* $n \geq k > 0$ *and* β *is* k*-good. Then there are arbitrarily large* $k-1$*-good ordinals below* β.

Proof. Let $\gamma < \beta$. Pick some $\alpha > \beta$ realizing $tp_{n,k}(\gamma, \beta)$. The facts that $\gamma < \beta < \alpha$ and β is $k-1$-good can be expressed in the language $\mathcal{L}'_{n,k}$. So the statement "$\exists y(\gamma < y < x) \wedge (y$ is $(k-1$-good)" belongs to $tp_{n,k}(\gamma, \alpha) = tp_{n,k}(\gamma, \beta)$. Hence, there is $\delta, \gamma < \delta < \beta$ which is $k-1$-good.
□

Let us now define a refinement of the forcing $\mathcal{P}$.

Definition 3.15 The set $\mathcal{P}^*$ is the subset of $\mathcal{P}$ consisting of sequences $p = \langle p_n \mid n < \omega \rangle$ so that for every n, $\ell(p) \le n < \omega$ and $\beta \in \mathrm{dom}(a_n)$ there is a nondecreasing converging to infinity sequence of natural numbers $\langle k_m \mid n \le m < \omega \rangle$ so that for every $m \ge n$ $a_m(\beta)$ is k_m-good, where $p_m = \langle a_m, A_m, f_m \rangle$.

The orders on $\mathcal{P}^*$ are just the restrictions of $\le$ and $\le^*$ of $\mathcal{P}$. The following lemma is crucial for showing the Prikry property of $\langle \mathcal{P}^*, \le, \le^* \rangle$.

Lemma 3.16 *$\langle \mathcal{P}^*, \le^* \rangle$ is κ_0-closed.*

Proof. Let $\langle p^\alpha \mid \alpha < \mu < \kappa_0 \rangle$ be a $\le^*$-increasing sequence of elements of $\mathcal{P}^*$. Suppose for simplicity that $\ell(p^0) = 0$ and hence for every $\alpha < \mu$ $\ell(p^\alpha) = 0$. Let $p^\alpha_n = \langle a^\alpha_n, A^\alpha_n, f^\alpha_n \rangle$ for every $n < \omega$ and $\alpha < \mu$. For each $n < \omega$ set $f_n = \bigcup_{\alpha<\mu} f^\alpha_n$ and $a_n = \bigcup_{\alpha<\mu} a^\alpha_n$. Let β be a $\sup(\mathrm{dom}(\bigcup_{n<\omega} a_n))$. We like to extend a_n by corresponding to β an ordinal $\delta_n < \lambda_n$ which is above $\cup(\mathrm{rng}(a_n))$, E_n-above every element of $\mathrm{rng}(a_n)$ and also is n-good. Such δ_n exists by Lemma 3.13. Set $b_n = a_n \bigcup \{\langle \beta, \delta_n \rangle\}$ and $B_n = \bigcap_{\alpha<\mu} \pi^{-1''}_{\delta_n \max(\mathrm{rng}(a_\alpha))}(A^\alpha_n)$. We define $q_n = \langle b_n, B_n, f_n \rangle$ and $q = \langle q_n \mid n < \omega \rangle$. Then $q \ge^* p^\alpha$, for every $\alpha < \mu$ and $q \in \mathcal{P}^*$. Since the only new element added is β and for every $n < \omega$ $b_n(\beta) = \delta_n$ is n-good. □

Now it is routine to show the Prikry condition for $\langle \mathcal{P}^*, \le, \le^* \rangle$.

Lemma 3.17 *$\langle \mathcal{P}^*, \le, \le^* \rangle$ satisfies the Prikry condition.*

Lemma 3.18 *For every $n < \omega$ $\mathcal{P}^* \simeq \mathcal{P}^* \restriction n \times \mathcal{P}^* \backslash n$.*

Lemma 3.19 *$\langle \mathcal{P}^*, \le \rangle$ does not add new bounded subsets to κ and it adds λ new ω-sequences to κ.*

Unfortunately, $\mathcal{P}^*$ still collapses κ^{++} to κ^+.
Let us now define an equivalence relation on $\mathcal{P}^*$.

Definition 3.20 Let $p = \langle p_n \mid n < \omega \rangle$, $q = \langle q_n \mid n < \omega \rangle \in \mathcal{P}^*$. We call p and q equivalent and denote this by $p \leftrightarrow q$ iff

(1) $\ell(p) = \ell(q)$

(2) for every $n < \ell(p)$ $p_n = q_n$

(3) there is a nondecreasing sequence $\langle k_n \mid \ell(p) \leq n < \omega\rangle$ with $\lim_{n\to\infty} k_n = \infty$ and $k_{\ell(p)} > 2$ such that for every n, $\ell(p) \leq n < \omega$ the following holds:

(a) $f_n = g_n$

(b) $\mathrm{dom}(a_n) = \mathrm{dom}(b_n)$

(c) $\mathrm{rng}(a_n)$ and $\mathrm{rng}(b_n)$ are realizing the same k_n-type, (i.e. the least ordinals coding $\mathrm{rng}\, a_n$ and $\mathrm{rng}\, b_n$ are such)

(d) $A_n = B_n$,
where $p_n = \langle a_n, A_n, f_n\rangle$ and $q_n = \langle b_n, B_n, g_n\rangle$.
Notice that, in particular the following is also true:

(e) for every $\delta \in \mathrm{dom}(a_n) = \mathrm{dom}(b_n)$ $a_n(\delta)$ and $b_n(\delta)$ are realizing the same k_n-type

(f) for every $\delta \in \mathrm{dom}(a_n) = \mathrm{dom}(b_n)$ and $\ell \leq k_n$ $a_n(\delta)$ is ℓ-good if $b_n(\delta)$ is ℓ-good

(g) for every $\delta \in \mathrm{dom}(a_n) = \mathrm{dom}(b_n)$ $\max(\mathrm{rng}(a_n))$ projects to $a_n(\delta)$ the same way as $\max(\mathrm{rng}(b_n))$ projects to $b_n(\delta)$, i.e. the projection functions $\pi_{\max(\mathrm{rng}(a_n)),a_n(\delta)}$ and $\pi_{\max(\mathrm{rng}(b_n)),b_n(\delta)}$ are the same.

Let us also define a preordering $\to$ on $\mathcal{P}^*$.

Definition 3.21 Let $p, q \in \mathcal{P}^*$.
Set $p \to q$ iff there is a sequence of conditions $\langle r_k \mid k < m < \omega\rangle$ so that

(1) $r_0 = p$

(2) $r_{m-1} = q$

(3) for every $k < m - 1$

$$r_k \leq r_{k+1} \quad \text{or} \quad r_k \leftrightarrow r_{k+1}\ .$$

The next two lemmas show that $\langle \mathcal{P}^*, \to\rangle$ is a nice subforcing of $\langle \mathcal{P}^*, \leq\rangle$.

Lemma 3.22 *Let $p, q, s \in \mathcal{P}^*$. Suppose that $p \leftrightarrow q$ and $s \geq p$. Then there are $s' \geq s$ and $t \geq q$ such that $s' \leftrightarrow t$.*

Proof. Let $\langle k_n \mid \ell(p) = \ell(q) \leq n < \omega\rangle$ be as in 3.20(3) witnessing $p \leftrightarrow q$. We need to define $s' = \langle s'_n \mid n < \omega\rangle$ and $t = \langle t_n \mid n < \omega\rangle$. Set $s'_n = t_n = s_n$ for every $n < \ell(p) = \ell(q)$. Set also $s'_n = s_n$ for every $n < \ell(s)$. Now let $\ell(p) \leq n < \ell(s)$. We show that $q_n = \langle b_n, B_n, g_n\rangle$ extends to s_n in the ordering of Q_n and then we'll set $t_n = s_n$. Let $p_n = \langle a_n, A_n, f_n \rangle$. By 2.16(3), $f_n = g_n$ and $A_n = B_n$.

We know that $s_n \geq \langle a_n, A_n, f_n \rangle$ (in the ordering of Q_n), hence $s_n(\max(\mathrm{dom}(a_n))) \in A_n$ and for every $\beta \in \mathrm{dom}(a_n)$ $s_n(\beta) = \pi_{\max(\mathrm{rng}(a_n)), a_n(\beta)}(s_n(\max(\mathrm{dom}(a_n))))$. But by 3.20(3)

$$\pi_{\max(\mathrm{rng}(a_n)), a_n(\beta)} = \pi_{\max(\mathrm{rng}(b_n)), b_n(\beta)}$$

and $\mathrm{dom}(a_n) = \mathrm{dom}(b_n)$. Thus, $s_n \geq \langle b_n, A_n, f_n \rangle = q_n$.

Suppose now that $n \geq \ell(s)$. Let $p_n = \langle a_n, A_n, f_n \rangle$, $q_n = \langle b_n, A_n, f_n \rangle$ and $s_n = \langle c_n, C_n, h_n \rangle$.

Case 1

$k_n = 3$.
Then we first extend s_n to a condition $s'_n \in Q_{n1}$ and proceed as above.

Case 2

$k_n > 3$.
Set $s'_n = s_n$. Then $\mathrm{rng}\, a_n$ and $\mathrm{rng}\, b_n$ are realizing the same k_n-type.

Thus it is possible to find $\widetilde{d}_n$ realizing the same $k_n - 1$-type over $\mathrm{rng}\, b_n$ as $\mathrm{rng}\, c_n$ over $\mathrm{rng}\, a_n$. Let d_n be the order preserving function from dom a_n onto $\widetilde{d}_n$. Set $t_n = \langle d_n, C_n, h_n \rangle$.

This completes the construction. $s' = \langle s'_n \mid n < \omega \rangle$ and $t = \langle t_n \mid n < \omega \rangle$ are as desired.
□

Lemma 3.23 *For every $p, q \in \mathcal{P}^*$ such that $p \to q$ there is $s \geq p$ so that $q \to s$.*

Proof. The proof is an inductive application of the previous lemma. Thus, suppose for example that

$$\begin{array}{ccc} q & \leftrightarrow & c \\ & & \vee\! | \\ a & \leftrightarrow & b \\ \vee\! | & & \\ p & & \end{array}$$

i.e. a, b, c are witnessing $p \to q$. We apply Lemma 3.22 to a, b and c. It provides equivalent $c' \geq c$ and $a' \geq a$. But then $a' \geq p$ and $q \to a'$, since

$$\begin{array}{ccc} a' & \leftrightarrow & c' \\ & & \vee\vert \\ q & \leftrightarrow & c \end{array}$$

Lemma 3.24 *$\langle \mathcal{P}^*, \to \rangle$ satisfies λ-c.c.*

Proof. Let $\langle p^\alpha \mid \alpha < \lambda \rangle$ be a sequence of elements of $\mathcal{P}^*$. Using the Δ-system argument it is easy to find a stationary $S \subseteq \lambda$, $\delta < \min(S)$, $\ell < \omega$ so that for every $\alpha, \beta \in S$, $\alpha < \beta$ the following holds

(a) $\ell(p^\alpha) = \ell$

(b) for every $n < \ell$ p^α_n and p^β_n are compatible

(c) for every $n \geq \ell$ let $p^\alpha_n = \langle a^\alpha_n, A^\alpha_n, f^\alpha_n \rangle$, then

  (i) $A^\alpha_n = A^\beta_n$

  (ii) f^α_n, f^β_n are compatible and $\min(\mathrm{dom}(f^\beta_n) \setminus \delta) \geq \beta > \sup(\mathrm{dom}(f^\alpha_n)) + \sup(\mathrm{dom}(a^\alpha_n))$

  (iii) $a^\alpha_n \restriction \delta = a^\beta_n \restriction \delta$

  (iv) $\min(\mathrm{dom}(a^\beta_n) \setminus \delta) \geq \beta > \sup(\mathrm{dom}(f^\alpha_n)) + \sup(\mathrm{dom}(a^\alpha_n))$

  (v) $\mathrm{rng}(a^\alpha_n) = \mathrm{rng}(a^\beta_n)$.

Let $\alpha < \beta$ be in S. We claim that p^α and p^β are compatible in $\langle \mathcal{P}^*, \to \rangle$. Define equivalent conditions $p \geq p^\alpha$ and $q \geq p^\beta$. First we set $p_n = q_n = p^\alpha_n \cup p^\beta_n$ for $n < \ell$. Let $\tau^\alpha = \min\left(\bigcup_{n \geq \ell} \mathrm{dom}(a^\alpha_n) \setminus \delta \right)$ and $\tau^\beta = \min\left(\bigcup_{n \geq \ell} \mathrm{dom}(a^\alpha_n) \setminus \delta \right)$. Assume for simplicity that $\tau^\alpha \in \mathrm{dom}(a^\alpha_\ell)$ and $\tau^\beta \in \mathrm{dom}(a^\beta_\ell)$. By 3.15 there is a nondecreasing converging to infinity sequences of natural numbers $\langle k_m \mid \ell \leq m < \omega \rangle$ so that for every $m \geq \ell$ $a^\alpha_m(\tau^\alpha) = a^\beta_m(\tau^\beta)$ is k_m-good. Let $n \geq \ell$.

Case 1

$k_n \leq 4$.
Pick some $\nu \in A^\alpha_n = A^\beta_n$. Set $p_n = q_n = f^\alpha_n \cup f^\beta_n \cup \{\langle \gamma, \pi_{\max(\mathrm{rng}(a^\alpha_n)), a^\alpha_n(\gamma)}(\nu) \rangle \mid \gamma \in \mathrm{dom}(a^\alpha_n)\} \cup \{\langle \gamma, \pi_{\max(\mathrm{rng}(a^\beta_n)), a^\beta_n(\gamma)}(\nu) \mid \gamma \in \mathrm{dom}(a^\beta_n)\}$. The condition (c) above insures that this is a function in Q_{n1}.

Case 2

$k_n > 4$.
Using Lemmas 3.13 and 3.14 for $a_n^\alpha(\tau^\alpha) = a_n^\beta(\tau^\beta)$, we find t^α realizing the same $k_n - 1$-type over $\mathrm{rng}(a_n^\alpha \restriction \delta) = \mathrm{rng}(a_n^\beta \restriction \delta)$ as $\mathrm{rng}(a_n^\alpha \backslash \delta) = \mathrm{rng}(a_n^\beta \backslash \delta)$ does so that $\min(t^\alpha) > \max(\mathrm{rng}(a_n^\alpha))$. Set $a'_n = \mathrm{rng}(a_n^\alpha) \cup t^\alpha$. $\mathrm{rng}(a_n^\beta \backslash \delta)$ realizes over $\mathrm{rng}(a_n^\alpha \restriction \delta)$ the same type as t^α. Hence there is t^β so that

$$\min(\mathrm{rng}(a_n^\beta \backslash \delta)) = a_n^\beta(\tau^\beta) > \max(t^\beta)$$

and if $b'_n = \mathrm{rng}(a_n^\beta) \cup t^\beta$, then a'_n and b'_n are realizing the same $k_n - 1$-type. Now pick n-good ordinal ξ coding a'_n. Using the $k_n - 1$ equivalence of a'_n, b'_n find $k_n - 2$-good ordinal ρ coding b'_n and so that ξ and ρ (and hence also $a'_n \cup \{\xi\}$ and $b'_n \cup \{\rho\}$ realize the same $k_n - 2$-type. Pick some $\gamma > \bigcup_{k<\omega} \left(\mathrm{dom}(f_k^\beta) \cup \mathrm{dom}(a_k^\beta) \right)$. Let a_n be the order isomorphism between $\mathrm{dom}(a_n^\alpha) \cup \mathrm{dom}(a_n^\beta) \cup \{\gamma\}$ and $a'_n \cup \{\xi\}$. Let b_n be the order isomorphism between $\mathrm{dom}(a_n^\alpha) \cup \mathrm{dom}(a_n^\beta) \cup \{\gamma\}$ and $b'_n \cup \{\rho\}$. We define $p_n = \langle a_n, A_n^\alpha, f_n^\alpha \cup f_n^\beta \rangle$ and $q_n = \langle b_n, A_n^\alpha, f_n^\alpha \cup f_n^\beta \rangle$.

By the construction such defined p and q are equivalent. So we are done.
□

The forcing with $\langle \mathcal{P}^*, \rightarrow \rangle$ preserves the cardinals, does not add new bounded subsets to κ and makes $2^\kappa = \kappa^{++}$.

References

1. T. Jech, *Set Theory*, 3rd edition, Springer Monographs in Mathematics (Springer, 2006).
2. A. Kanamori, *The Higher Infinite. Large Cardinals in Set Theory from their Beginnings*, Perspectives in Mathematical Logic (Springer-Verlag, Berlin, 1994); xxiv+536 pp.

THE TURING DEGREES: AN INTRODUCTION

Richard A. Shore*
Department of Mathematics
Cornell University
Ithaca NY 14853 USA
shore@math.cornell.edu

We introduce some of the basic techniques of recursion theory as used to analyze the structure of the Turing degrees, i.e. the structure of relative complexity of computation. We begin with the classical Kleene-Post constructions. These are extended to the primary technique developed here: forcing in arithmetic. We also study some domination properties and their relation to the degrees. The structural results we cover include embedding theorems both for partial orders and lattices. At the global level, we apply these results and techniques to prove that the theory of the degrees is recursively equivalent to that of full second order arithmetic. The theory of the degrees below $0'$ is equivalent to that of true first order arithmetic. Finally, we show that a relation on the degrees below $0'$ that is invariant under the double jump is definable in that degree structure if and only if it is definable in first order arithmetic.

Contents

---

*Partially supported by NSF Grants DMS-0852811 and DMS-1161175 and as a Visiting Professor in the Department of Mathematics and the Institute for Mathematical Sciences at the National University of Singapore with partial funding from the John Templeton Foundation. These lecture notes are based on a short course given at the Institute as part of the Computational Prospects of Infinity II: AII Graduate Summer School (15 June – 13 July 2011). Earlier versions were based on our courses at Cornell and especially one in the fall of 2007. That semester all the students took turns taking notes and one, Mia Minnes, took notes for essentially all the lectures. These notes formed the basis for an early draft for much of the material presented here and more. We thank all our students and especially those from that semester.

1. Introduction

Our goal in these lectures is to explore the fundamental notion of effective computability (recursiveness) and, more specifically that of relative complexity of computation (relative recursiveness). The formal definitions that best captures the intuitions, first, that some function (or set) is computable and, second, that one set (or function) is easier to compute than another are those of Turing. We work with the natural numbers $\mathbb{N}$ and subsets of and functions on them. Turing machines supply a formalism for describing what are generally agreed to be all the intuitively computable functions and the basic notion of general computability of one set (or function) from another. While there were many other formalisms introduced in an attempt to capture these notions we now know that they are all equivalent and we can simply think of the programs in any general purpose computer language as supplying our basic list of such functions. To describe the notion of computing one set from another we equip our (Turing) machines with an "oracle". For $A, B \subseteq \mathbb{N}$, we say that A is recursive in (or (Turing) computable from) B, $A \leq_T B$, if, when we want to decide if $n \in A$, we allow our basic machines at any point in their computation to generate an $m \in \mathbb{N}$, ask if $m \in B$ and receive the correct answer from the oracle for B. The machine may then continue on with its computation. We say that A and B are (Turing) equivalent, $A \equiv_T B$, if $A \leq_T B$ and $B \leq_T A$.

This notion of relative recursiveness (computability) defines a symmetric, transitive relation on the subsets of (or functions on) $\mathbb{N}$. As usual, we move to the equivalence classes of this relation which are called the (Turing) degrees. The degree of a set A, $\deg(A)$, is then $\{B|B \equiv_T A\}$, often denoted by $\mathbf{a}$. These degrees then form a partial order under the induced ordering $\mathbf{a} \leq \mathbf{b}$. (Note that we can pass between sets A and functions f by using graphs of functions $(\{\langle x,y\rangle\,|f(x)=y\})$ in one direction and characteristic functions of sets ($C_A(n)=1$ if $n \in A$ and $C_A(n)=0$ if $n \notin A$) in the other. We generally abuse notation and confuse sets and functions in this way. It is a basic fact (or an exercise to check) that these procedures preserve Turing degree.) We denote the structure of these degrees and partial ordering by $\mathcal{D}$. It is our primary object of study in these lectures.

It is easy to see that this partial order has a least element $\mathbf{0}$ the degree of the empty set 0 (or equivalently of any recursive set, i.e. one computable by a Turing machine). It also has join operator $\mathbf{a} \vee \mathbf{b} = \deg(A \oplus B)$ where $A \oplus B = \{2n|n \in A\} \cup \{\{2n+1|n \in B\}$. (It is an exercise to see that this defines the least upper bound of $\mathbf{a}$ and $\mathbf{b}$ in $\mathcal{D}$.) A deeper fact about the ordering is that it has the countable predecessor property, i.e. $\{\mathbf{b}|\mathbf{b} < \mathbf{a}\}$ is at most countable for any degree $\mathbf{a}$. The point here is that there is computable listing of the Turing machines (which have "space" for an oracle) and so of the functions they compute Φ_e (Φ_e^A when relative to the oracle A). Thus $\{B|B \leq_T A\}$ is countable for every set A and so, a fortiori, $\{\mathbf{b}|\mathbf{b} < \mathbf{a}\}$ is at most countable. One of our major goals is to see what more we can say about this ordering in first order or algebraic terms. Is it a linear ordering? It is an uppersemilattice (usl) but is it a lattice? What orderings can be embedded into it, etc.?

There are also important and remarkable connections between relative computability as expressed in structural properties of $\mathcal{D}$ and approximations to, and growth rates of, functions on the one hand and definability in arithmetic on the other. This story begins with the halting problem and its generalization, the (Turing) jump to all sets and degrees. The halting problem is traditionally defined as $0' = \{e|\Phi_e(e) \text{ converges}\}$ with degree $\mathbf{0}'$. Its generalization is given by $A' = \{e|\Phi_e^A(e) \text{ converges}\}$ with degree $\mathbf{a}'$. (Again it is a basic fact (or an exercise to see) that this operation is well defined on degrees. The fact that it is strictly increasing is essentially the classical result on the undecidability of the halting problem but relative to arbitrary oracles.) In terms of definability in arithmetic, A' is essentially the same as the set of existential formulas true in $\mathbb{N}$. (It is certainly of the same degree as this set but even more closely related to it.) For A', the corresponding set is that of the existential formulas in arithmetic with an added unary relation for A. Iterations of this operator move up the levels of quantifier complexity. (See Theorem 1.10.) As for approximations, the sets computable from A' are precisely those with approximations recursive in A. (See Theorem 1.11.) The connections to rates of growth are a bit more subtle but quite important. (See Sec. 5.) Thus another important concern in these lectures will the jump operator and its relation to the order structure on D. In particular, in parallel with our study of $\mathcal{D}$, we will extensively study the structure of the degrees recursive in the halting problem, $\mathcal{D}(\leq \mathbf{0}')$.

Finally, in addition to investigating the algebraic or first order properties of these structures we will analyze their second order or metamathematical properties. For example, we will characterize the complexity (in terms of Turing degree and more) of their theories, $Th(\mathcal{D})$ and $Th(\mathcal{D}(\leq \mathbf{0}')$, the sets of sentences true in these structures as well as study the sets and relations definable in them.

We give a brief list of some of the notations, conventions and basic results that are used later in Sec. 1.1. We begin our main task of analyzing $\mathcal{D}$ and $\mathcal{D}(\leq \mathbf{0}')$ in Sec. 2. There we introduce the idea of dividing up a complex property into simpler ones (called requirements) and the method of approximating the sets we want to build having the desired property by finite initial segments. In a construction by such approximations we want to satisfy the requirements in terms of these approximations in such a way that we guarantee the sets constructed have the desired properties. These ideas all come from the seminal paper on degree theory by Kleene and Post [1954]. In hindsight, these constructions can be seen as simple examples of Cohen's later method of forcing but implemented in the setting of arithmetic instead of set theory. In the rest of these lectures, we formalize and develop a more general approach to forcing in recursion theory. We then apply it to prove most of our results about the structures $\mathcal{D}$ and $\mathcal{D}(\leq \mathbf{0}')$, both mathematical and metamathematical.

We do not attempt to give a historical account of the material presented in these lectures. Indeed, most of the proofs are not the original ones. However, we do give, in Notes at the end of most subsections, basic attributions and references for most of the results to provide some historical perspective.

1.1. *Some background material*

We hope that almost all of the material in this subsection is already known to the readers. If so, it can be skipped, If not, it can be taken on faith, worked out as exercises or found in the first couple of sections of any standard text.

We begin with a few facts about Turing computations and how the basic programs Φ_e work with oracles.

Definition 1.1: There is master (universal) recursive function,

$$\varphi(\sigma, e, x, s) = y$$

where the variables are σ a finite binary string (initial segment of a characteristic function or set), e a number (index), x a number (input), s a number (steps of the computation). The expression means that the Turing machine with index e and oracle restricted to σ given input x and run for s many steps converges and outputs y.

Conventions: If the computation asks question outside the domain of σ or does not converge in s steps we announce that the computation is divergent.

Properties:

(i) Use: If $\sigma \subseteq \tau$ and $\varphi(\sigma, e, x, s) \downarrow= y$ then $\varphi(\tau, e, x, s) = y$

(ii) Permanence: If $s < t$ and $\varphi(\sigma, e, x, s) \downarrow= y$ then $\varphi(\sigma, e, x, t) \downarrow= y$

(iv) The domain of φ is computable, in other words there is a procedure to decide whether φ converges on any given tuple (σ, e, x, s). This procedure simply runs the machine with index e on input x and oracle σ. If the machine arrives at an output by step s, then answer yes (and otherwise, answer no).

Definition 1.2: (Computations from Oracles) $\Phi_e^A(x) = y$ means that $\exists \sigma \subseteq A \exists s \big[\varphi(\sigma, e, x, s) \downarrow= y\big]$. So $\Phi_e^A(x)$ is a partial function (recursive in A). We define the *use of a computation* $\Phi_e^A(x) = y$ as the least n such that $\varphi(A \restriction n+1, e, x, s) = y$. We also say that $\sigma = A \restriction n$ is the *axiom* (about the oracle A) that gives this computation. Note that if A is changed at or below the use then this axiom no longer applies and we no longer have the same computation giving the output y.

Definition 1.3: We adopt two additional conventions when the oracle is a finite string σ. First, we run the Turing machine for only $|\sigma|$, the length of σ, many steps so we write $\Phi_e^\sigma(x)$ for $\Phi_{e,|\sigma|}^\sigma(x)$. Second, we require that for $\Phi_e^\sigma(x)$ to converge we must have $x < |\sigma|$. (Roughly speaking we must read the input before giving an output.)

Definition 1.4: (Turing Reducibility) $A \leq_T B$ means $\exists e(\Phi_e^B = A)$. $A \equiv_T B$ means that $A \leq_T B$ and $B \leq_T A$. The equivalence classes under this relation are the (Turing) degrees **a** and **b** (of A and B, respectively). They are ordered by the induced partial order, $\mathbf{a} \leq \mathbf{b}$.

Intuitively this means that there is a Turing machine with oracle B that computes A.

Exercise 1.5: Turing reducibility is symmetric and transitive.

Definition 1.6: A set is recursively enumerable (r.e.) in A if it is the domain of a partial function recursive in A, i.e. of some Φ_e^A.

Exercise 1.7: For any sets A and B, $A \leq_T B$ if and only if both A and $\bar{A}$ are r.e. in B.

The archetypic r.e. in A set is its jump A'.

Definition 1.8: (Jump Operator) The jump of A, A', is $\{e|\Phi_e^A(e)\downarrow\}$. The iterations of this operator are defined by $A^{(n+1)} = (A^{(n)})'$. (We use $\downarrow$ to stand for "converges".)

Exercise 1.9: A' is r.e. in A. Moreover, the jump operator is order preserving and hence well defined on the degrees, i.e., if $A \leq_T B$ then $A' \leq_T B'$. In addition, $A <_T A'$ for every set A.

We assume some standard language for first order arithmetic containing for example the functions $+$ and $\times$, the relation $\leq$ and the constants 0 and 1 (or also an additional unary predicate for a set A). The standard syntactic hierarchy of Σ_n (or Σ_n^A) and Π_n (Π_n^B) formulas in prenex normal form are defined by counting the number of alternations or quantifiers as usual. Typically one includes bounded quantification $\exists x < s$ and $\forall x < s$ in the matrix of these formulas. One can instead add the master function φ of Definition 1.1 into the language. There are normal forms for these formulas that show that, for example, one can assume that there is only one quantifier of each sort as the types of the quantifiers at the beginning of the formula alternate. The primary connection between the classes of sets defined by such formulas which are also denoted by Σ_n^A and Π_n^B are given by the hierarchy theorem. (We say that a set is Δ_n^A if it is both Σ_n^A and Π_n^A.)

Theorem 1.10: ***(Post's Hierarchy Theorem)***

(1) $B \in \Sigma_{n+1}^A \Leftrightarrow B$ *is RE in some* Π_n^A *set.*

(2) $A^{(n)}$ is Σ_n^A m-complete for $n > 0$,i.e. for any $B \in \Sigma_n^A$ there is a recursive function f such that $\forall n(n \in B \Leftrightarrow f(n) \in A^{(n)})$. This is even stronger than the assertion that $B \leq_T A$.

(3) $B \in \Sigma_{n+1}^A \Leftrightarrow B$ is r.e. in $A^{(n)}$.

(4) $B \in \Delta_{n+1}^A \Leftrightarrow B \leq_T A^{(n)}$.

There is an important connection between the Δ_2^B sets (which are those recursive in B') by clause 4 of this theorem and those with approximations computable in B:

Theorem 1.11: ***(Shoenfield Limit Lemma)*** *$A \leq_T B' \Leftrightarrow \exists f \leq_T B$ such that $\forall x\big(A(x) = \lim_{s\to\infty} f(x,s)\big)$. Note that asserting that $\lim_{s\to\infty} f(x,s)$ exists means that $f(x,s)$ is eventually constant for each x.*

The jump and its iterations are important markers along the highway of complexity for sets. Thus we will often take some construction and ask where along this road the sets or degrees constructed lie or can be made to lie, such as below $\mathbf{0}'$ or $\mathbf{0}''$ or some other $\mathbf{0}^{(n)}$. Another measure of complexity is where the jump(s) of the set constructed lie. For example, we might ask if $A' \equiv_T 0'$ (the smallest possible value) or if $A'' \equiv_T 0''$ (the largest possible value for any r.e. set or one recursive in $0'$). These ideas are captured in the definition of the jump hierarchy and the generalized jump hierarchy.

Definition 1.12: For $n \geq 1$, $X \in GL_n$ if and only if $X^{(n)} \equiv_T (X \vee 0')^{(n-1)}$ (by convention, $Z^{(0)} = Z$ for every Z); $X \in GH_n$ if and only if $X^{(n)} \equiv_T (X \vee 0')^{(n)}$. If $X \leq_T 0'$ then these conditions simplify and we say that $X \in L_n$ if $X^{(n)} \equiv_T 0^{(n)}$ and $X \in H_n$ if $X^{(n)} \equiv_T 0^{(n+1)}$. We indicate the corresponding degree classes by boldfacing: $\mathbf{GL}_n, \mathbf{GH}_n, \mathbf{L}_n$ and $\mathbf{H}_n$. The complementary classes are indicted by $\overline{\mathbf{GL}}_n, \overline{\mathbf{GH}}_n, \bar{\mathbf{L}}_n$ and $\bar{\mathbf{H}}_n$ where the last two refer to the complement within the degrees below $\mathbf{0}'$. These notations are read as (generalized) low$_n$ or (generalized) high$_n$.

Notes: For basic background including the material of this section we recommend the classics texts on recursion theory Rogers [1987] and Soare [1987] or the more encyclopedic Odifreddi [1989] and [1999].

2. Embeddings into the Turing Degrees

2.1. *Embedding partial orders in $\mathcal{D}$*

Based on only the background information on the Turing degrees mentioned in the Introduction, we know only that $\mathcal{D}$ is an uppersemilattice of size $2^{\aleph_0}$ (Exercise) with least element and the countable predecessor property. It also has an operator, the Turing jump, which is strictly increasing and closely related to the quantifier complexity of the definitions of sets and functions in arithmetic. The only specific degrees we know are $\mathbf{0}$ and the iterations of the jump beginning with $\mathbf{0}'$. Are there others? Is $\mathcal{D}$ a linear order? If not, how "wide" is it? How far away from being a linear order? Where do these other degrees lie with respect to the ones we already know? We begin answering these questions by considering what is perhaps the simplest question and showing that $\mathcal{D}$ is not a linear order.

Notation 2.1: We write $A|_T B$, A is Turing incomparable with B, for $A \not\leq_T B$ *and* $B \not\leq_T A$.

Theorem 2.2: ***(Kleene and Post)*** $\exists A_0, A_1 (A_0|_T A_1)$.

How can we approach such a result. We recast the desired properties of the sets we want to construct into a list of simpler ones R_e called requirements. Then we choose an approximation procedure so that we can build a sequence of approximations $\alpha_{i,s}$ "converging" to A_i such that the information in an approximation $\langle\alpha_{i,s}\rangle$ can be sufficient to guarantee that we satisfy one of the requirements in the sense that R_e is true of any pair $A_i \supset \alpha_{i,s}$.

Proof: We build A_0, A_1. The requirements necessary to guarantee the theorem are:

$$R_{\langle e,j\rangle} : \Phi_e^{A_j} \neq A_{1-j}$$

for all $e \in \mathbb{N}$, $j \in \{0, 1\}$. It is clear that if the sets we construct satisfy each requirement then the sets satisfy the demands of the theorem. Our approximations in this case are finite binary strings (so initial segments of characteristic functions) $\alpha_{j,s}$ such that $A_j = \cup_s \alpha_{j,s}$.

The construction cannot be recursive because A_0, A_1 cannot both be recursive and incomparable. But, the approximations will not change once defined at some x; in other words, $\alpha_{j,s} \subseteq \alpha_{j,s+1}$ so we get better and better approximations.

What actions satisfy a requirement? Given $\alpha_{j,s}$ ($j = 0, 1$), we want $\alpha_{j,s+1} \supseteq \alpha_{j,s}$ to guarantee that we satisfy $R_{\langle e,j \rangle}$. For definiteness, let $j = 0$. We want $\alpha_0 \supseteq \alpha_{0,s}$, $\alpha_1 \supseteq \alpha_{1,s}$ such that for any $A_0 \supseteq \alpha_0$, $A_1 \supseteq \alpha_1$, $\Phi_e^{A_0} \neq A_1$. In other words,

$$\exists x \neg (\Phi_e^{A_0}(x) = A_1(x)).$$

We can choose x as the first place x at which $\alpha_{1,s}$ is not defined (formally $x = \text{dom}(\alpha_{1,s}) = |\alpha_{1,s}|$). Ask if $\exists \alpha_0 \supseteq \alpha_{0,s}(\Phi_e^{\alpha_0}(x) \downarrow)$. If so, we can choose the *"least"* such α_0. To which ordering does the "least" refer here? We make a master list of all convergent computations $\varphi(\sigma, e, x, t)$, i.e. $\{\langle \sigma, e, x, t \rangle : \varphi(\sigma, e, x, t) \downarrow\}$ where we write $\varphi(\sigma, e, x, t)$ or $\Phi_{e,t}^{\sigma}(x)$ to mean the result of running the eth Turing machine on input x for t many steps with oracle questions answered by the finite string σ (which must be long enough to answer them) and then *least* refers to the least quadruple $\langle \sigma, e, x, s \rangle$ in this list. From now on we, usually without comment, use "least" in this sense of being the first object enumerated in some given search.

Then, we set $\alpha_{0,s+1} = \alpha_0$ and $\alpha_{1,s+1} = \alpha_{1,s}^{\wedge}(1 - \Phi_e^{\alpha_0}(x))$. By the standard properties of Turing machines, if $A_0 \supseteq \alpha_0 = \alpha_{0,s+1}$ and $A_1 \supseteq \alpha_{1,s+1}$ then

$$\Phi_e^{A_0}(x) = \Phi_e^{\alpha_0}(x) \neq 1 - \Phi_e^{\alpha_0}(x) = A_1(x).$$

What if no such α_0 exists? We do nothing, i.e. we set $\alpha_{i,s+1} = \alpha_{i,s}$. This finishes the construction.

A general principle of our constructions is do the best you can, and if you cannot do anything useful, then do nothing and hope for the best (i.e. that what you can is enough). In this case, it *is* enough because if $A_0 \supseteq \alpha_{0,s}$ then $\Phi_e^{A_0}(x) \uparrow$. (If $\Phi_e^{A}(x) \downarrow$ for any $A \supseteq \alpha_{0,s}$ then the computation only requires finitely much information about A and so $\Phi_e^{\alpha}(x) \downarrow$ for some finite initial segment α of A. As $A_0 \supseteq \alpha_{0,s}$ we can certainly take this α to extend $\alpha_{0,s}$ as well if $\Phi_e^{A_0}(x) \downarrow$.) So $\Phi_e^{A_0}$ is not total and can certainly then not be the characteristic function of a set, i.e. $\Phi_e^{A_0} \neq A_1$.

Thus we have actually verified that the construction satisfies all the requirements and so provides the desired sets: Consider $R_{\langle e,j \rangle}$. Look at the stage s at which we acted for this requirement. Either we did something (defined $\alpha_{i,s+1} \neq \alpha_{i,s}$) which guaranteed the requirement by guaranteeing that $\Phi_e^{A_j}(x) \downarrow \neq A_{1-j}(x)$ at some x; or we did nothing by setting $\alpha_{i,s+1} = \alpha_{i,s}$ but in that case we also guaranteed that the requirement is satisfied by making $\Phi_e^{A_j}(x) \uparrow$ for some x. □

Question 2.3: How do we know that this construction keeps going, i.e. that there is no point after which we always "do nothing". If that were the case, then both A_0 and A_1 would be finite, so certainly not Turing incomparable. Why does this not happen? Is it necessary to include another requirement to guarantee this: $Q_e : \alpha_{j,s} \geq e$? (These would be easy to satisfy.) Whenever we do act on a requirement, we make one of the α's longer and since infinitely often there is an index e which does not look at its oracle and outputs 0, at the stage at which we deal with the requirement with index e, we automatically extend the approximation. Hence, both strings are extended infinitely often. This is a common phenomenon. Constructions often do more than one expects.

Question 2.4: How complicated are A_0 and A_1? We want a bound on their complexity such as $A_0, A_1 \leq_T 0^{(n)}$ (this would also give definability properties). To determine what n is, let us look back at the construction. By recursion, we have $\alpha_{j,s}$. To calculate $\alpha_{j,s+1}$, we asked one question:

$$\exists \alpha_0 \supseteq \alpha_{0,s}(\Phi_e^{\alpha_0}(x)\downarrow)?$$

This is a Σ_1 question so $0'$ can answer it and tell us which case to implement. The "do nothing" case is easy to do. For the other case, we have to enumerate the master list $\{\langle \sigma, e, x, t\rangle : \varphi(\sigma, e, x, t)\downarrow\}$, which we can do effectively. So, once $0'$ told us which case we're in, everything else is recursive. Hence, $A_0, A_1 \leq 0'$.

Question 2.5: Where do A_0, A_1 lie in the jump hierarchy? Because of the symmetry of the construction, even though $A_0 \not\equiv_T A_1$, they should have some of the same properties. Are they low (or can we add something to the construction to make sure that they are low)? (Recall: A_0 is low iff $A_0' \leq_T 0'$ iff $\{e : \Phi_e^{A_0}(e)\downarrow\} \leq_T 0'$.)

We can add a new requirement:

$$N_{e,j} : \text{make } \Phi_e^{A_j}(e)\downarrow \text{ if we can.}$$

Suppose that at stage s we are acting on $N_{e,0}$. We have $\alpha_{j,s}$ and ask if

$$\exists \alpha_0 \supseteq \alpha_{0,s}(\Phi_e^{\alpha_0}(e)\downarrow)?$$

If the answer is yes, let $\alpha_{0,s+1}$ be the least such α_0 and let $\alpha_{1,s+1} = \alpha_{1,s}$. On the other hand, if the answer is no, then do nothing and put $\alpha_{j,s+1} = \alpha_{j,s}$ This is called *deciding or forcing the jump.* The terminology will be better understood after Sec. 3.2.

Claim 1: The construction is still recursive in $0'$: Our actions for requirements $P_{e,j}$ are the same as before. For $N_{e,j}$, $0'$ can decide if $\exists\alpha_0 \supseteq \alpha_{0,s}(\Phi_e^{\alpha_0}(e)\downarrow)$.

Claim 2: We can compute A_0' from $0'$. Since the whole construction is recursive in $0'$, $0'$ can go along the construction until it gets to the stage s at which we act for $N_{e,0}$. Then, it sees what the construction does and can compute A_0' from this action.

Claim 3: We can relativize the construction to any degree $\mathbf{x}$ to get incomparables A_j^X between X, X' such that $(A_j^X)' = X'$. By relativizing, we mean that at each part of the computation where we have oracle α_j, we instead have the oracle $X \oplus \alpha_j$. At the end, we build $X \oplus A_j$. The verification of the construction goes through as before.

Claim 4: It is easy to extend the construction to more than two incomparables. We can change the requirements to

$$P_{e,i,j} : \Phi_e^{A_j} \neq A_i \qquad i \neq j.$$

Thus, we can produce countably many low pairwise incomparables between 0 and $0'$, indeed all with jumps uniformly recursive in $\mathbf{0}'$.

Exercise 2.6: Show that the sets A_i of the original construction (for Theorem 2.2) are already low.

Notation 2.7: Given any sequence $\langle A_i | i \in I\rangle$ of sets we let $\oplus\{A_i | i \in I\} = \{\langle i, x\rangle \,| i \in I \,\&\, x \in A_i\}$. Conversely, given any set A we let $A^{[i]}$ denote the set $\{\langle i, x\rangle \mid \langle i, x\rangle \in A\}$. We let $A^{[\hat{i}]} = \oplus\{A_j | i \neq j\} = \{\langle j, x\rangle \,| i \neq j \;\&\; x \in A_j\}$.

In general, given a countable partial order $\mathcal{P}$, can we embed it in $\mathcal{D}$ or in $\mathcal{D}(\leq 0')$ or in the low degrees? Let $\mathcal{P} = \{p_0, p_1, \ldots\}, \leq_{\mathcal{P}}$. Without loss of generality, we can assume that p_0 is the least element of $\mathcal{P}$. (If $\mathcal{P}$ does not have a least element, add one in and then any embedding of this enlarged partial order gives an embedding of the original $\mathcal{P}$.) We build A_i such that $A_i \leq_T A_j$ if and only if $p_i \leq_{\mathcal{P}} p_j$. To do so, we build C_i and let $A_j = \oplus\{C_i : p_i \leq_{\mathcal{P}} p_j\}$. Does $p_i \leq_{\mathcal{P}} p_j$ imply that $A_i \leq_T A_j$? By transitivity,

$$\langle k, x\rangle \in A_i \;\Leftrightarrow\; x \in C_k \wedge p_k \leq_{\mathcal{P}} p_i \;\Rightarrow\; \langle k, x\rangle \in A_j \;\Leftrightarrow\; x \in C_k \wedge p_k \leq_{\mathcal{P}} p_j$$

so if $\leq_{\mathcal{P}}$ is recursive, $i \leq_{\mathcal{P}} j$ implies that $A_i \leq_T A_j$. We can use this fact to embed recursive partial orders in the low degrees by using the construction above to guarantee incomparability when needed and the recursiveness of $\mathcal{P}$ with this simple argument to guarantee comparability when needed. If a partial order is not recursive, it is at least recursive in some oracle so relativizing the proof for recursive partial orders gives an embedding into $\mathcal{D}$. Perhaps this is the best we can do – it may not intuitively obvious that $\mathcal{D}(\leq 0')$ is a universal countable partial order. We begin by constructing a recursive universal partial order. The construction is an example of the method of finite approximations being used to build sets with properties not necessarily expressed in terms of Turing degrees. We then embed it into $\mathcal{D}(\leq \mathbf{0}')$.

Theorem 2.8: *There is a recursive universal partial order $\mathcal{P}$, i.e. one such that every countable partial order $\mathcal{Q}$ can be embedded in $\mathcal{P}$.*

Proof: We build $\mathcal{P}$ by finite approximations, $\mathcal{P} = \cup \mathcal{P}_s$. At stage s we have a finite partial order $\mathcal{P}_s$ and extend it to $\mathcal{P}_{s+1}$ such that for every subset of $\mathcal{P}_s$, every one element partial order extension is realized in $\mathcal{P}_{s+1}$. That is, for every subset $M \subset P_s$, and a particular partial order relation on $M \cup \{z\}$ (z, a new element), add z to $\mathcal{P}$ and define its relation to the elements in $P_s \setminus M$ as dictated by the axioms of partial orders. Thus we can prove that given any partial order and any finite subset and any extension by one element, there is a new partial order that realizes that extension. We can apply this finitely many times to take care of each finite subset and each possible one-element partial order extension. This construction is recursive so we have a recursive partial order.

To see that $\mathcal{P}$ is universal, consider any countable partial order $\mathcal{Q}$. We use a forth argument to embed $\mathcal{Q}$ into $\mathcal{P}$. That is, if $\mathcal{Q} = \{q_0, q_1, \ldots\}$ we define the embedding f by recursion. Start with $f(q_0) = p_0$ and then, given $f(q_m)$ for $m < n$, define $f(q_n)$ to be an element of $\mathcal{P}$ realizing (up to this finite isomorphism) the same extension of $\{f(q_m) | m < n\}$ that q_n does of $\{q_0, \ldots, q_{n-1}\}$. □

Proposition 2.9: *Every recursive partial order $\mathcal{P} = (P, \leq_{\mathcal{P}})$ can be embedded in $\mathcal{D}$.*

Proof: Let p_i enumerate the elements of P. We build sets C_i and let $A_i = \oplus\{C_j : p_j \leq_{\mathcal{P}} p_i\}$ so if $p_k \leq_{\mathcal{P}} p_j$ then $A_k \leq_T A_j$ since $\leq_{\mathcal{P}}$ is recursive.

Requirements: $R_{k,j,e} : p_k \nleq_{\mathcal{P}} p_j$ implies $A_k \nleq_T A_j$ i.e. $\forall e \Phi_e^{C_j} \neq C_k$.

Approximations: Finitely many finite binary strings $\gamma_{j,s}$. We set $C_j = \cup\gamma_{j,s}$. Then we approximate the A_i by

$$A_{i,s} = \oplus\{\gamma_{j,s} : p_j \leq_{\mathcal{P}} p_i\}$$

i.e. $A_{i,s}$ is defined at $\langle j, x\rangle$ if $\gamma_{j,s}(x)$ is defined. Think of each $\gamma_{j,s}$ as a finite partial function and $A_{i,s}$ is the sum of these partial functions. To make A_i a total characteristic function we set $A_i(\langle j, x\rangle (x) = 0$ if $p_j \not\leq_{\mathcal{P}} p_i$.

Suppose we wish to act for $R_{k,j,e}$ at stage $s = \langle k, j, e\rangle$. We have $A_{j,s}, A_{k,s}$ finite characteristic functions determined by the $\gamma_{i,s}$ so far defined. To guarantee that $\Phi_e^{A_j} \neq A_k$, could we take $x = |\gamma_{k,s}|$ and ask if there is extension of the γ's such that $\Phi_e^{A_j}(x) \downarrow$ to diagonalize? The problem is that an extension of the γ's which guarantees convergence might also determine the value $A_k(x)$, so we might not be able to diagonalize.

To make x not interfere with the computation from A_j, we want an $x = \langle n, y\rangle$ such that $p_n \not\leq_{\mathcal{P}} p_j$. Also, to be able to define A_k at x, we need $p_n \leq p_k$ (otherwise the relevant column is always empty). We also need $\langle n, y\rangle \geq |\gamma_{k,s}|$. So we want $p_n \not\leq p_j$ and $p_n \leq_{\mathcal{P}} p_k$. By assumption, $p_k \not\leq_{\mathcal{P}} p_j$, so choose $n = k$. Then let $x = \langle k, |\gamma_{k,s}|\rangle$.

Now, ask for least extension of the γ's which makes $\Phi_e^{A_j}(x) \downarrow$. This only depends on γ_i for $p_i \leq_{\mathcal{P}} p_j$. If such an extension exists, put $A_k(x) = 1 - \Phi_e^{A_j}(x)$. If there is no such extension, do nothing. Then, go to stage $s + 1$.

To verify that the construction satisfies all the requirements, for $R_{k,j,e}$ consider the stage $s = \langle k, j, e\rangle$. Either we extended some γ or we did not. If we extended some γ, then there is x such that $\Phi_e^{A_j}(x) \downarrow \neq A_k(x)$. If we did not, then no such extension exists and since A_j extends $\gamma_{j,s}$, $\Phi_e^{A_j}(x) \uparrow$. □

Corollary 2.10: *Every countable partial order can be embedded in $\mathcal{D}$.*

Corollary 2.11: *The one-quantifier theory of $(\mathcal{D}, \leq_T)$ is decidable.*

Proof: A one-quantifier existential sentence is equivalent to a disjunction of ones of the form

$$\varphi \equiv \exists x_1 \exists x_2 \cdots \exists x_n \big(x_i \leq x_j \wedge \cdots \wedge x_j \not\leq x_k \wedge \cdots \wedge x_n = x_n\big).$$

(Note that if we can decide whether an existential sentence is true or false then we can flip the answers to decide if universal sentences are true or false.)

Given such a disjunct, we ask if there is a partial order that satisfies one of the disjuncts. If not, then $(\mathcal{D}, \leq_T)$ cannot because it itself is a partial order. So suppose $(\mathcal{P}, \leq_{\mathcal{P}}) \models \mathcal{P}$. If we can embed $\mathcal{P}$ into $\mathcal{D}$ then we are done because embedding preserves atomic sentences. Not every partial ordering can be embedded into $\mathcal{D}$ (for example, huge ones cannot). But if there is any partial order that satisfies φ then there is a finite partial order that satisfies it, because φ only mentions n elements. So, we can assume that $\mathcal{P}$ is finite, hence recursive. Then, the theorem above says that $\mathcal{P}$ embeds into $\mathcal{D}$. The last piece of the proof is to verify that we can answer the question of whether φ is satisfiable by a partial order. Well, we can enumerate all partial orders of size at most n and then check each one. And, if φ is satisfiable by a partial order then it is satisfiable by a member of the list. □

Exercise 2.12: If the recursive partial order $\mathcal{P}$ of Proposition 2.9 has a least element 0, then embedding f into $\mathcal{D}$ can be chosen such that $f(0) = \mathbf{0}$. Then Corollary 2.10 can be extended to partial orders with least element and Corollary 2.11 to the language with a constant for $\mathbf{0}$.

Question 2.13: We ask the following questions about the proof of embedding theorem, Proposition 2.9:

(1) How complicated are the images of the partial order under the embedding? We claim that $A_i \leq_T 0'$ uniformly. Indeed the whole construction and so the C_i are (uniformly) recursive in $0'$. To compute $A_i(x)$ where $x = \langle j, n \rangle$ we first ask if $p_j \leq p_i$ (the partial ordering is recursive). If not, $A_i(x) = 0$. If so, we can follow the construction recursively in $0'$ until it is decided if $x \in C_j$.
(2) Can we ensure that all the A_i are low? We can add requirements

$$N_e : \text{ make } \Phi_e^{\oplus A_i}(e) \downarrow \text{ if we can.}$$

To act for N_e still takes just a $0'$ question. Alternatively, instead of adding infinitely many requirements we can add a top element 1 to $\mathcal{P}$. The construction then gives $A_1 = \oplus C_j \leq 0'$ and we can then just make sure that A_1 is low.

Corollary 2.14: *Every countable partial older can be embedded in $\mathcal{D}(\leq \mathbf{0}')$ and so its one quantifier theory is decidable.*

An alternative approach to these results begins with strengthened versions of incomparability.

Definition 2.15: The set $\{A_i : i \in \mathbb{N}\}$ is *independent* if no A_i is computable from the join of finitely many of the other A_j. The set $\{A_i : i \in \mathbb{N}\}$ is *very independent* if $A_i \not\leq_T \oplus_{j \neq i} A_j$ for all i.

Very independent implies independent because $A_{i_1} \oplus \cdots \oplus A_{i_n} \leq_T \oplus_{j \neq i} A_j$ if no $i_k = i$ ($x \in A_i \Leftrightarrow \langle i, x\rangle \in \oplus_{j \neq i} A_j$). However, while independence is a degree theoretic notion, very independence is not. This is proved in the following exercises.

Exercise 2.16: Find $\{A_i : i \in \mathbb{N}\}$ very independent.

Exercise 2.17: Find $\{A_i : i \in \mathbb{N}\}, \{B_i : i \in \mathbb{N}\}$ such that $\{A_i : i \in \mathbb{N}\}$ is very independent, $\{A_i : i \in \mathbb{N}\}$ is not, but $A_i \equiv_T B_i$.

Definition 2.18: An *uppersemilattice (usl)* is a partially ordered set $\mathcal{P}$ such that every pair of elements x, y in $\mathcal{P}$, has a least upper bound, $x \vee y$.

Exercise 2.19: Every usl $\mathcal{L}$ is locally countable, i.e. for any finite $F \subset L$ the subusl $\mathcal{F}$ of $\mathcal{L}$ generated by F (i.e. the smallest one containing F) is finite. Moreover, there is a uniform recursive bound on $|\mathcal{F}|$ that depends only on $|F|$.

Exercise 2.20: Given finite usls $Q \subset P$ and an usl extension $\hat{Q}$ of Q generated over Q by one new element (with $\hat{Q} \cap P = Q$), prove that there is an usl extension $\hat{P}$ of P containing $\hat{Q}$.

Exercise 2.21: Prove that there is a recursive usl $\mathcal{L}$ such that every countable usl can be embedded in it (as an usl).

Exercise 2.22: Every countable usl $\mathcal{L}$ can be embedded in $\mathcal{D}$ and even in $\mathcal{D}(\leq \mathbf{0}')$ (preserving $\vee$ as well as $\leq$). Hint: Use a very independent set C_i. If $\mathcal{L} = \{l_i\}$ send l_i to $\oplus\{C_j | l_j \not\geq l_i\}$.

Notes: The finite extension method for constructing degrees was developed in Kleene and Post [1954]. It was the seminal paper on the structure of the Turing degrees. They proved, among others, Theorem 2.2, the existence of a countable family of independent sets and Proposition 2.9 for finite partial orders and that these theorems are true in the degrees below $\mathbf{0}'$. Sacks [1961] and [1963] contain Corollary 2.10 and much more. Corollary 2.11 is pointed out in Lerman [1972].

We will see in Theorem 3.44 that every countable lattice can be embedded in $\mathcal{D}$ but not by the methods used here in the sense that there is no countable lattice $\mathcal{L}$ which is countably universal, let alone a recursive one. Indeed local finiteness fails and there are $2^{\aleph_0}$ many lattices generated by four elements. We provide such with seven generators in Sec. 3.4.

What about uncountable partial orders, usls and lattices? Of course, they must have the countable predecessor property, i.e. $\{y|y \leq x\}$ is countable for every x. Sacks [1961] shows that all partial orders of size $\aleph_1$ with the countable predecessor property can be embedded into $\mathcal{D}$. For lattices this follows from Abraham and Shore [1986] where the embedding is made onto an initial segment of $\mathcal{D}$. Sacks [1961] shows that all those with the countable successor property can be embedded. However, it is consistent that $2^{\aleph_0} = \aleph_2$ and there is an usl of size $\aleph_2$ with the countable predecessor property which cannot be embedded in $\mathcal{D}$ (Groszek and Slaman [1983]). It is a long standing open question if every partial order of size $2^{\aleph_0}$ with the countable predecessor property can be embedded in $\mathcal{D}$ (Sacks [1963]).

2.2. *Extensions of embeddings*

We now look at extensions of embedding results which give information about the 2-quantifier theory of $(\mathcal{D}, \leq_T)$.

Theorem 2.23: *(**Avoiding Cones**) For every $A > 0$ there is B such that $A|_T B$.*

Proof: Given a set A, we build B such that $A \not\leq_T B$, $B \not\leq_T A$. There are two kinds of requirements:

$$P_e : \Phi_e^A \neq B \qquad\qquad Q_e : \Phi_e^B \neq A.$$

The construction is by finite binary string approximations β_s for B. At the end, we let $B = \cup_s \beta_s$.

Suppose at stage s we work to satisfy P_e. We have β_s and construct β_{s+1} guaranteeing that B meets the requirement. We ask for the value of $\Phi_e^A(|\beta_s|)$. If $\Phi_e^A(|\beta_s|) \uparrow$ then P_e is satisfied so do nothing. Otherwise, put $\beta_{s+1} = \beta_s\hat{\ }(1 - \Phi_e^A(|\beta_s|))$. So, $B(|\beta_s|) = \beta_{s+1}(|\beta_s|) \neq \Phi_e^A(|\beta_s|)$. Observe that at this stage we ask a question that A' can answer and then carry out a recursive procedure.

Likewise, suppose at stage s we work to satisfy Q_e. We ask if there is an x and an extension σ of β_s such that $\Phi_e^\sigma(x) \downarrow\neq A(x)$. If no such extension exists, do nothing. If there is such an extension, let β_{s+1} be the least such extension. Note that this is a Σ_1^A question followed by a recursive procedure, so this step is recursive in A'.

To verify that this construction works, observe that all the P_e are clearly satisfied. Suppose we fail to satisfy Q_e. Then at stage s there was no x and $\sigma \supset \beta_s$ such that $\Phi_e^\sigma(x) \downarrow \neq A(x)$. If $\Phi_e^B(x) \uparrow$ for any x then Q_e is satisfied. Otherwise, we claim that A is recursive: To compute $A(x)$, look for a $\sigma \supseteq \beta_s$ such that $\Phi_e^\sigma(x) \downarrow$. There is one since $\Phi_e^B(x) \downarrow$. The value computed with oracle σ must be $A(x)$. This contradicts our assumption that A is not recursive. Thus, Q_e is satisfied. □

Exercise 2.24: The B of Theorem 2.23 can be made recursive in A' and indeed we can guarantee (or the construction already does) that $B' \equiv_T A'$.

Exercise 2.25: Every maximal chain (i.e. linearly ordered subset) in $\mathcal{D}$ is uncountable.

Exercise 2.26: For every countable set of nonrecursive degrees there is a degree incomparable with each of them.

Exercise 2.27: Every nonempty maximal antichain (i.e. pairwise incomparables) in $\mathcal{D}$ is uncountable.

Exercise 2.28: Every maximal independent set of degrees is uncountable.

Theorem 2.29: ***(Minimal Pair)*** *There are $A, B > 0$ such that $A \wedge B = 0$. In other words, for all C, if $C \leq_T A, B$ then $C \equiv_T 0$.*

Proof: We build A, B by finite approximations α_s, β_s. There are three kinds of requirements:

$P_e : \Phi_e \neq B, \quad Q_e : \Phi_e \neq A \quad$ and $\quad N_{e,i} : \Phi_e^A = \Phi_i^B = C \Rightarrow C$ is recursive.

To satisfy P_e, Q_e (respectively): given α_s (β_s), ask if $\Phi_e(|\alpha_s|) \uparrow$ (or $\Phi_e(|\beta_s|) \uparrow$). If yes, then the requirement is already satisfied so let $\alpha_{s+1}(|\alpha_s|) = 0$ ($\beta_{s+1}(|\beta_s|) = 0$). Otherwise, let $\alpha_{s+1}(|\alpha_s|) = 1 - \Phi_e(|\alpha_s|)$ ($\beta_{s+1}(|\beta_s|) = 1 - \Phi_e(|\beta_s|)$).

Suppose at stage s we work on $N_{e,i}$. Ask if $(\exists \alpha \supseteq \alpha_s)$ $(\exists \beta \supseteq \beta_s) \exists x (\Phi_e^\alpha(x) \downarrow \neq \Phi_i^\beta(x) \downarrow)$. If such extensions exist, pick the first pair (α, β) which satisfies the condition and put $\alpha_{s+1} = \alpha$, $\beta_{s+1} = \beta$. If no such extensions exist, do nothing.

To verify that the construction works, first notice that all the P_e and Q_e are satisfied so $A, B > 0$. For $N_{e,i}$, we may assume that $\Phi_e^A = \Phi_i^B = C$ as otherwise the requirement is automatically satisfied. We want to show that C is recursive. Consider α_s, β_s for the stage s at which we work on

$N_{e,i}$. To compute $C(x)$, find any finite extension $\alpha \supseteq \alpha_s$ such that $\Phi_e^\alpha(x)$. (There is one since $A \supseteq \alpha_s$ and $\Phi_e^A(x)\downarrow$.) We claim that $\Phi_e^\alpha(x) = C(x)$. If not, there is a $\beta \supseteq \beta_s$ with $\beta \subseteq B$ such that $\Phi_e^\beta(x) = \Phi_e^B(x) = C(x)$ and so we would have acted at s with α and β contrary to our assumption. □

We frequently use the idea seen in this proof of searching for extensions that give different outputs when used as oracles for a fixed Φ_e and, if we find them, doing some kind of diagonalization. If there are none, we generally argue that Φ_e^A is recursive (or recursive in the relevant notion of extension as in Theorem 2.33). We extract the appropriate notion and provide some terminology.

Definition 2.30: We say that two strings σ and τ *e-split* (or form an *e-splitting*) if $\exists x(\Phi_e^\sigma(x)\downarrow \neq \Phi_e^\tau(x)\downarrow)$. We denote this relation by $\sigma|_e\tau$ and say that σ *and* τ *e-split at* x. Note that by our conventions in Definition 1.3, $\Phi_e^\sigma(x) = \Phi_{e,|\sigma|}^\sigma(x)$ is a recursive relation as is $\exists x(\Phi_e^\sigma(x)\downarrow \neq \Phi_e^\tau(x)\downarrow)$, i.e. $\sigma|_e\tau$.

Exercise 2.31: We may make the A and B of Theorem 2.29 low or note that as constructed they are already low. We can also relativize the result: $\forall C \exists A, B(A \wedge B \equiv C \ \&\ A' \equiv B' \equiv C')$.

We want a notion similar to minimal pairs but with an arbitrary countable ideal of degrees playing the role of $\mathbf{0}$.

Definition 2.32: $\mathcal{C} \subseteq \mathcal{D}$ is an *ideal* in the uppersemilattice $\mathcal{D}$ if it is closed under joins and is closed downwards (i.e. if $\mathbf{y} \in \mathcal{C}$ and $\mathbf{x} \leq \mathbf{y}$ then $\mathbf{x} \in \mathcal{C}$).

Theorem 2.33: *(**Exact Pair**) If $\mathcal{C}$ is any countable ideal in $\mathcal{D}$, there are* $\mathbf{a}, \mathbf{b}$ *such that* $\mathcal{C} = \{\mathbf{x} : \mathbf{x} \leq_T \mathbf{a}, \mathbf{b}\} = \{\mathbf{x} : \mathbf{x} \leq_T \mathbf{a}\} \cap \{\mathbf{x} : \mathbf{x} \leq_T \mathbf{b}\}$.

An alternative statement of the theorem is the following:

Theorem 2.34: *If $C_1 \leq_T C_2 \leq_T \cdots$ is an ascending sequence, then there are A, B such that* $\{X : X \leq_T A, B\} = \{X : \exists n(X \leq_T C_n)\}$.

Exercise 2.35: These two statements are equivalent. We can list all the sets D_j with degrees in a countable ideal $\mathcal{C}$ and then consider the ascending sequence $C_i = \oplus_{j<i} D_j$.

We prove the second formulation of the theorem.

Proof: Given $\langle C_n \rangle$ ascending in Turing degree, we build A and B such that

- for all n, $C_n \leq_T A, B$ and
- $C \leq_T A, B$ implies that $C \leq_T C_n$ for some n.

Therefore, we need to satisfy the requirements

$$R_n : C_n \leq_T A, B \qquad N_{e,i} : \Phi_e^A = \Phi_i^B = C \Rightarrow \exists n(C \leq_T C_n).$$

We build A and B by approximations α_s, β_s. Instead of these being finite strings, however, they are matrices. In each matrix, finitely many columns are entirely determined and there is finitely much additional information. Suppose at stage s we work for R_n. Choose the first column in each of α_s, β_s which has no specifications as yet. Let α_{s+1} (β_{s+1}) be the result of putting C_n into that column of α_s (β_s) and leaving the rest of the approximation unchanged. This action is computable in C_n. Otherwise, suppose at stage s we work to satisfy $N_{e,i}$. Ask if $\exists x(\exists \alpha \supseteq \alpha_s)$ $(\exists \beta \supseteq \beta_s)(\Phi_e^\alpha(x)\downarrow = \Phi_i^\beta(x)\downarrow)$ with the domains of α and β being only finitely larger than those of α_s and β_s, respectively. If such extensions exist, let $(\alpha_{s+1}, \beta_{s+1})$ be the least such pair of extensions. If no such extensions exist, do nothing.

Now, A and B meet the condition that $C_n \leq_T A, B$ for all n because all the R_n requirements are satisfied. Consider the stage s at which we deal with requirement $N_{e,i}$. We may assume that $\Phi_e^A = \Phi_i^B = C$ as otherwise the requirement is automatically satisfied. We want to prove $C \leq_T C_n$ for some n. Indeed let n be the largest m such that we have coded C_m into A and B by stage s. To compute $C(x)$, find any finite extension α of α_s such that $\Phi_e^\alpha(x)\downarrow$. (There is one since $A \supseteq \alpha_s$ and $\Phi_e^A(x)\downarrow$.) We claim that $\Phi_e^\alpha(x) = C(x)$. If not, there is a finite extension β of β_s with $\beta \subseteq B$ such that $\Phi_e^\beta(x) = \Phi_e^B(x) = C(x)$ and so we would have acted at s with α and β contrary to our assumption. The crucial point now is that checking whether $\alpha \supseteq \alpha_s$ is recursive in C_n. □

Corollary 2.36: *$\mathcal{D}$ is not a lattice.*

Proof: Let C_i be strictly ascending in Turing degree. (Such exist, for example, by Theorem 2.9.) Now let A and B be as in Theorem 2.34. If there were a C whose degree is the infimum of those of A and B then $C \leq_T A, B$ and so $C \leq_T C_n$ for some n. In this case, $C <_T C_{n+1} \leq_T A, B$ for a contradiction. □

Exercise 2.37: What is a bound on the complexity (degrees) of the A and B of Theorem 2.34 in terms of the C_n? Does $(\oplus C_n)'$ work? How about a better bound? How low can we make this bound? Consider also the special case that $C_n = 0^{(n)}$.

Exercise 2.38: Use the results of the previous exercise and Corollary 2.14 to show that $\mathcal{D}(\leq \mathbf{0}')$ is not a lattice.

Exercise 2.39: (**Extensions of Embeddings**) Given a finite usl $\mathcal{P}$ and a finite partial ordering $\mathcal{Q}$ extending $\mathcal{P}$ with no $x \in Q - P$ below any $y \in P$ and an usl embedding $f : \mathcal{P} \to \mathcal{D}$ prove that there is an extension g of f embedding $\mathcal{Q}$ into $\mathcal{D}$ as a partial order.

Notes: Theorems 2.23 and 2.29 and Corollary 2.36 are due to Kleene and Post [1954]. Exercises 2.26 and 2.27 to Shoenfield [1960]. Sacks [1961] proves Exercise 2.28 but Groszek and Slaman [1983] shows that it is consistent that $2^{\aleph_0} = \aleph_2$ but there is a maximal independent set of size $\aleph_1$. Theorem 2.33 and Exercise 2.38 are due to Spector [1956].

3. Forcing in Arithmetic and Recursion Theory

3.1. *Notions of forcing and genericity*

Forcing provides a common language for, and generalization of, the techniques we have developed in Sec. 2. It captures the idea of approximation to a desired object and how individual approximations guarantee (force) that the object we are building satisfies some requirement. Now approximations usually come with some sense of when one is better, or gives more information, than another. Of course, an approximation may have improvements which are incompatible with each other, i.e. the set of approximations is partially ordered. The intuition is that $p \leq q$ means that p refines, extends or has more information than q. We are generally thinking that the conditions are approximations to some object $G : \mathbb{N} \to \mathbb{N}$ (typically a set) and that if $p \leq q$ then the approximation p gives more information than q and so the class of potential objects that have p as an approximation is smaller then the one for q. In addition, we have some notion of what, at least at a basic level, the approximation p says about G. We formalize these ideas as follows:

Definition 3.1: A *notion of forcing* is a partial order $\mathcal{P}$ with domain a set P and binary relation $\leq_{\mathcal{P}}$. We call an element of $\mathcal{P}$ a *(forcing) condition.* For convenience, we assume that the partial order has a greatest element **1**. (For further restrictions see Definition 3.11.)

Example 3.2: If the notion of forcing is $(2^{<\omega}, \supseteq)$, then $\sigma \leq \tau \equiv \sigma \supseteq \tau$. In many of our previous constructions we used such binary strings σ as approximations to a set G such that $\sigma \subset G$. So the longer the string, the fewer sets that "satisfy" it, i.e. have it as an approximation (initial segment). This example is often called *Cohen forcing.*

Example 3.3: In Theorem 2.34, we used partial characteristic functions α defined on some finite set of columns and some finitely many additional points. Again we were approximating a set $G \supset \alpha$.

Example 3.4: If the notion of forcing is the set of perfect (i.e. every node has two incomparable extensions) recursive binary trees under $\subseteq$ then $S \leq T \equiv S \subseteq T$. (Here trees T are simply sets of finite strings, i.e. subsets of $\omega^{<\omega}$, which are downward closed, i.e. if $\tau \subseteq \sigma \in T$ then $\tau \in T$.) Think of such a tree T as approximating the set $[T]$ of its paths, i.e. $[T] = \{f | \forall n(f \upharpoonright n \in T\}$, so more information means fewer paths, i.e. more information about which path is being approximated. This notion of forcing is often called *Spector forcing* (or *perfect forcing* or *Sacks forcing* or other names for different variations).

What object is it exactly or what class of objects is it that a condition p approximates? For Cohen forcing a condition (string) σ approximates the class of sets $\{G | G \supset \sigma\}$. So the collection of all approximations to a single set G is simply $\{\sigma | \sigma \subset G\}$, the class of all the initial segments of G. We want to isolate the salient features of this set of conditions or any set $\mathcal{G} \subseteq \mathcal{P}$ that might be considered as an object its members are approximating. The general approach that we want for an arbitrary notion of forcing begins with that of a filter.

Rather than simply comparing any two elements, the idea is to compare each of them with the imaginary end point that we're approximating. That is, between two given positions and end goal, there is an element extending both of the given ones.

Definition 3.5: Two elements p, q are *compatible* if and only if $\exists r(r \leq p \wedge r \leq q)$. If p, q are incompatible we write $p \perp q$ (as opposed to *incomparable* which is written as $p \mid q$ to denote that $p \nleq q$ and $q \nleq p$).

Definition 3.6: $\mathcal{F} \subseteq \mathcal{P}$ is a filter on $\mathcal{P}$ if and only if $\mathcal{F}$ is nonempty, upward closed and for every $p, q \in \mathcal{F}$ there is an $r \in \mathcal{F}$ with $r \leq_{\mathcal{P}} p, q$.

We are thinking of filters as connected with the object we are approximating, the end goal.

Example 3.7: Suppose we want to approximate a set $G \in 2^\omega$ and our notion of forcing is $(2^{<\omega}, \supseteq)$ (finite binary strings). Then the set $\{\sigma : \sigma \subset G\}$ is a filter. In particular, the union of this set (filter) is the characteristic function G. It will commonly be the case that the object we want is defined from a filter by some "simple" operation such as union. We formalize this idea in Definition 3.11. Note that for finite strings, being comparable is the same as being compatible.

Example 3.8: Suppose we want to approximate a set $G \in 2^\omega$ and our notion of forcing is some countable set of infinite binary trees (not necessarily perfect) such as the recursive ones. Then the set $\{T : G \in [T]\} = \{T : \forall\sigma \subset G(\sigma \in T)\}$ is a filter: Suppose two trees both have G as a path. Then the tree which is the (set) intersection of the two trees is a common refinement. For upward closure, if G is a path on T and $T \subseteq S$ then G is also a path on S. In this case, the intersection of this filter is the characteristic function of G.

Suppose $\mathcal{F}$ is a filter on some notion of forcing $\mathcal{P}$. We can often associate some set or function with $\mathcal{F}$ in a canonical way. For example, for Cohen forcing we can naturally try $\cup\mathcal{F}$. For forcing with binary trees we might try $\cap\{[T] | T \in \mathcal{F}\}$. Does this always make sense even for Cohen or Spector forcing? For Cohen forcing it might be that $\cup\mathcal{F}$ is a finite string so itself a condition. For Spector forcing $\cap\{[T] | T \in \mathcal{F}\}$ could be a set of paths through a binary tree with more than one branch which might not necessarily be recursive or perfect. We need to add conditions on our filter to make sure we get a total function or a single set at the end. We might for example require for Cohen forcing that $\mathcal{F}$ contain strings of every (equivalently arbitrarily long) length, i.e. $(\forall n)(\exists\sigma \in \mathcal{F})(|\sigma| \geq n)$. For Spector forcing we could require that there are trees in $\mathcal{F}$ with arbitrarily long nodes σ before the first branching (i.e. σ has two immediate successors in the tree but no $\tau \subset \sigma$ does). To this end we add a function $V(p)$ representing the atomic information about our generic object determined by the condition p and the requirement that all generic filters meet certain dense sets defined in terms of V.

Definition 3.9: $D \subseteq \mathcal{P}$ is *dense in* $\mathcal{P}$ if $\forall p \in \mathcal{P} \exists q \in \mathcal{D}(q \leq_P p)$. D is *dense below* r if $\forall p \leq_P r \exists q \in \mathcal{D}(q \leq p)$.

In general we want the conditions guaranteeing (forcing) each of our requirements to be dense.

Definition 3.10: If $\mathcal{C}$ is a class of dense subsets of $\mathcal{P}$, we say that $\mathcal{G}$ is *$\mathcal{C}$-generic* if $\mathcal{G} \cap D \neq \emptyset$ for all $D \in \mathcal{C}$. We say that a sequence $\langle p_n \rangle$ of conditions is $\mathcal{C}$-generic if $\forall i(p_{i+1} \leq_{\mathcal{P}} p_i)$ and $\forall D \in \mathcal{C} \exists n(p_n \in D)$.

Definition 3.11: We always require that a notion of forcing have a *valuation function* $V : P \to \omega^{<\omega}$ which is recursive on $\mathcal{P}$ and continuous in the sense that if $p \leq_{\mathcal{P}} q$ then $V(p) \supseteq V(q)$. (We say that a partial recursive function φ is *recursive on a set* X if $X \subseteq \text{dom}(\varphi)$.) Moreover, we require that the sets $V_n = \{p| \ |V(q)| \geq n)\}$ are dense. We also require that any collection of dense sets that we consider for the construction of a generic filter or sequence include the V_n.

Example 3.12: In the Examples above we may define $V(p) = p$ for Cohen forcing. When the conditions are trees T, we may let $V(p)$ be the largest σ such that every $\tau \in T$ is comparable with σ. Show that the corresponding V_n are dense. What should V be for the forcing that constructs an exact pair?

Proposition 3.13: *If $\langle p_n \rangle$ is a $\mathcal{C}$-generic sequence then $\mathcal{G} = \{p|\exists n(p_n \leq p\}$ is a $\mathcal{C}$-generic filter containing each p_n.*

Proof: $\mathcal{G}$ is $\mathcal{C}$-generic because it contains an element, p_n, of D_n for all n. It is upward closed because if $p \in \mathcal{G}$ then $p \geq p_e$ for some e so if $q > p \geq p_e$ then $q \geq p_e$ as well. Finally, it is pairwise compatible because given $p \geq p_{e_1}$, $q \geq p_{e_2}$ then $p, q \geq p_e$ where $e = \max\{e_1, e_2\}$. □

If our collection of dense sets is countable then generic sequences and filters always exist.

Theorem 3.14: *If $\mathcal{C}$ is countable and $p \in \mathcal{P}$, then there is a $\mathcal{C}$-generic sequence $\langle p_n \rangle$ with $p_0 = p$ and so, by Proposition 3.13, a $\mathcal{C}$-generic filter $\mathcal{G}$ containing p.*

Proof: Let $\mathcal{C} = \{D_n|n \in \mathbb{N}\}$. We define $\langle p_n \rangle$ by recursion beginning with $p_0 = p$. If we have p_n then we choose any $q \leq p_n$ in D_n as p_{n+1}. One exists by the density of D_n. It is clear that $\langle p_n \rangle$ is a $\mathcal{C}$ generic sequence and so $\mathcal{G} = \{p|\exists n(p_n \leq p\}$ is $\mathcal{C}$-generic filter containing p. □

Exercise 3.15: Consider Cohen forcing. If $\mathcal{C}$ is countable (as it always is in our applications) and $\mathcal{G}$ is a $\mathcal{C}$-generic filter containing p, then there is a $\mathcal{C}$-generic sequence $\langle p_n \rangle \leq_T \mathcal{G}$ with $p_0 = p$ such that $\mathcal{G} = \{p|\exists n(p_n \leq p\}$.

Definition 3.16: We associate to each $\mathcal{C}$-generic sequence $\langle p_n \rangle$ or filter $\mathcal{G}$ the *generic function (or set)* $G = \cup V(p_n)$ or $\{V(p) | p \in \mathcal{F}\}$.

Proposition 3.17: *If G is associated with the $\mathcal{C}$-generic sequence $\langle p_n \rangle$ (filter $\mathcal{G}$) then $G \leq_T \langle p_n \rangle$ ($\mathcal{G}$).*

Proof: As V is recursive on $\mathcal{P}$ and the sets V_n of Definition 3.11 are included in $\mathcal{C}$, we can compute $G(n)$ by searching for a k such that $p_k \in V_n$ (or $p \in \mathcal{G}$) and then noting that $G(n) = V(p_k)$ $(V(p))$. □

As is our general practice, we often care about how hard it is to compute a $\mathcal{C}$-generic sequence, filter or function. We must begin with the complexity of $\mathcal{P}$ and then consider how hard it is to compute the generic sequence $\langle p_e \rangle$ and so the associated generic G. We view the elements of $\mathcal{P}$ as being (coded by) natural numbers. For convenience we let the natural number **1** be the greatest element of P.

Definition 3.18: A notion of forcing $\mathcal{P}$ is *A-recursive* (or **a**-*recursive*) if the set P and the relation $\leq_{\mathcal{P}}$ are recursive in A ($\in \mathbf{a}$). (As usual if $A = \emptyset$ ($\mathbf{a} = \mathbf{0}$) we omit it from the notation.) If $\mathcal{C} = \{C_n\}$ is a collection of dense sets in $\mathcal{P}$ then f is a *density function for* $\mathcal{C}$ if $\forall p \in P \forall n \in \mathbb{N}(f(p,n) \in C_n)$.

Proposition 3.19: *If $\mathcal{P}$ is an A-recursive notion of forcing and $\mathcal{C} = \{C_n\}$ is a uniformly A-recursive sequence of dense subsets of $\mathcal{P}$ and $p \in P$ then there is a $\mathcal{C}$-generic sequence $\langle p_n \rangle$ with $p_0 = p$ which is recursive in A. More generally, for an arbitrary notion of forcing $\mathcal{P}$, $p \in P$ and a class $\mathcal{C}$ of dense sets, if f is a density function for $\mathcal{C}$, then there is a $\mathcal{C}$-generic sequence $\langle p_n \rangle \leq_T f$ with $p_0 = p$. The generic G function associated with these filters or sequences are also recursive in A or f, respectively.*

Proof: If $\mathcal{P}$ is an A-recursive notion of forcing and $\mathcal{C} = \{C_n\}$ is a uniformly A-recursive sequence of dense subsets of $\mathcal{P}$, then we can define a density function $f \leq_T A$ by letting $f(p,n)$ be the least (in the natural order of $\mathbb{N}$) $q \leq_{\mathcal{P}} p$ with $q \in C_n$. The desired generic sequence is now given by setting $p_0 = p$ and $p_{n+1} = f(n, p_n)$. That G is recursive in A or f now follows from Proposition 3.17. □

Note that the generic filter $\mathcal{G}$ defined from the generic sequence $\langle p_n \rangle$ in Proposition 3.13 is Σ_1 in $\langle p_n \rangle$ but not necessarily recursive in it.

3.2. *The forcing language and deciding classes of sentences*

The ad hoc approach to constructions presented in Sec. 2 looks at the specific theorem we want to prove, decides what are the specific requirements we need to meet, and then builds the desired sets accordingly. For example, this is what we did to build $A|_T B$. Our approximations were $\mathcal{P} = \{\langle \alpha, \beta \rangle\}$. The requirements were $\Phi_e^A \neq B$ (and $\Phi_e^B \neq A$). Given α, β, we could find $\langle \hat{\alpha}, \hat{\beta} \rangle \leq \langle \alpha, \beta \rangle$ which would guarantee the requirement. In particular, if one exists, we chose a specific $\langle \hat{\alpha}, \hat{\beta} \rangle \leq \langle \alpha, \beta \rangle$ such that $\exists x \Phi_e^{\hat{\alpha}}(x) \downarrow \neq \hat{\beta}(x) \downarrow$; if not, we took $\langle \alpha, \beta \rangle$. In the terminology of forcing, we had dense sets $D_e =$

$$\{\langle \alpha, \beta \rangle : \exists x \Phi_e^{\alpha}(x) \downarrow \neq \beta(x) \downarrow \text{ or } (\forall \langle \hat{\alpha}, \hat{\beta} \rangle \leq \langle \alpha, \beta \rangle)(\neg[\exists x \Phi_e^{\hat{\alpha}}(x) \downarrow \neq \hat{\beta}(x) \downarrow])\}.$$

Likewise, we defined dense sets C_e, which guaranteed that $\Phi_e^B \neq A$. Then if $\mathcal{G}$ is $\{D_e, C_e\}$-generic, $G_0 \mid_T G_1$.

In this manner, each of the proofs we did earlier by constructions with requirements can be translated to dense sets and generics for the dense sets D_e determined by the conditions that guarantee (force) that we satisfy the eth requirement. However, the benefit of the forcing technology comes in the form of the generality it allows. For example, we could try to tackle many of the constructions at once. We need to define the forcing relation ($\Vdash$) more generally, by induction on formulas φ that somehow say that if $p \Vdash \varphi$ then $\varphi(G)$ holds for the set or function G determined by any sufficiently generic filter $\mathcal{G}$.

Thus we want a relation $\Vdash$ between conditions $p \in \mathcal{P}$ and sentences $\phi(\mathtt{G})$ (where we use $\mathtt{G}$ as the formal symbol that is to be interpreted as our generic set or function G). This relation should approximate truth in the sense just described. We could use a standard language of arithmetic (in set theoretic forcing, one would use the language of set theory) augmented with another parameter ($\mathtt{G}$) for the set we are building, and possibly other parameters ($\bar{F}$) for given sets or functions. For our purposes it is more convenient to use the master (universal) partial recursive predicates $\phi(G, \bar{F}, e, \bar{x})$ and the standard normal form theorems mentioned in Sec. 1.1 and described below.

We fix some finite sequence of functions or sets $\bar{F}$ and view them as fixed parameters that appear in our formulas. This allows us to formalize relativizations to such $\bar{F}$ as well as other notions. For the most part, however, we can ignore them in our proofs as the relativizations are almost always straightforward.

We use $\phi_n(G, \bar{F}, e, x_0, \ldots, x_{n-1})$ to mean that the eth Turing machine with oracles G and $\bar{F}$ (which we are viewing as a fixed (possibly empty) set or function parameters that depends on the notion of forcing and is included in the oracle) running for x_0 many steps on inputs $x_1, \ldots, x_{n-1}$ converges. Our conventions are that if the machine runs for s many steps then it must first read the program and inputs and then can ask about the value of any one of the oracles at n only after writing out n and must then read the answer. So, in particular, $\phi_n(G, \bar{F}, e, x_0, \ldots x_{n-1})$ can hold only if $e, x_1, \ldots, x_{n-1} < x_0$ and for any information $G(m)$ or $F(m)$ about the oracles used in the computation $m, G(m)$ and $F(m)$ are also less than x_0. Thus if $\phi_n(G, \bar{F}, e, x_0, \ldots x_{n-1})$ holds then it only depends on $G \upharpoonright x_0$ (and $\bar{F} \upharpoonright x_0$ although we ignore this fact as we are thinking of the $\bar{F}$ as given parameters) in the sense that it is also true for any $G' \supseteq G \upharpoonright x_0$ and. In this case we also say that $\phi(\sigma, \bar{F}, e, x_0, \ldots x_{n-1})$ holds for any $\sigma \supseteq G \upharpoonright x_0$. Thus, crucially, the predicates $\phi_n(\sigma, \bar{F}, e, x_0, \ldots x_{n-1})$ are uniformly recursive in $\bar{F}$. Moreover, every Σ_1 sentence about G and $\bar{F}$ is equivalent to one of the form $\exists x_0 \phi(G, \bar{F}, e, x_0)$ and so to $\exists x_0 \phi(G \upharpoonright x_0, \bar{F}, e, x_0)$. Every Π_1 sentence about G and $\bar{F}$ is equivalent to one of the form $\forall x_0 \neg\phi(G, \bar{F}, e, x_0)$ and so to $\forall x_0 \neg\phi(G \upharpoonright x_0, \bar{F}, e, x_0)$. More generally, for $n > 0$, every Σ_{2n+1} sentence about G and $\bar{F}$ is equivalent to one of the form $\exists x_{2n} \forall x_{2n-1} \ldots \exists x_0 \phi(G, \bar{F}, e, x_0, \ldots, x_{2n})$; every Σ_{2n} sentence about G and $\bar{F}$ is equivalent to one of the form $\exists x_{2n} \forall x_{2n-1} \ldots \forall x_0 \neg\phi(G, \bar{F}, e, x_0, \ldots, x_{2n})$ and similarly for Π_n sentences about G and $\bar{F}$. Thus, it suffices to consider only formulas beginning with a list of quantifiers of alternating type followed by a predicate of the form $\phi(G, \bar{F}, e, x_0, \ldots, x_{n-1})$ (if the final quantifier is $\exists$) or $\neg\phi(G, \bar{F}, e, x_0, \ldots, x_{n-1})$ (if the final quantifier is $\forall$). (Note that n may be larger than the number of quantifiers and we include constants m in our language for every $m \in \mathbb{N}$.)

Notation 3.20: We use $\neg\varphi$ to stand for the canonical equivalent of the negation of φ, i.e. change each quantifier ($\exists$ to $\forall$ and vice versa) and the matrix (ϕ to $\neg\phi$ and vice versa). So, in particular, $\neg\neg\varphi = \varphi$.

Notation 3.21: We use $\mathcal{G}$ for the generic filter, G for $\cup\{V(p) | p \in \mathcal{G}\}$, the set or function that we are building and $\mathtt{G}$ for the symbol in language that stands for that set or function.

We define the forcing relation $p \Vdash \varphi$ for $p \in \mathcal{P}$ and φ a sentence of our language by induction on the complexity of sentences. The usual definition in standard languages proceeds by induction on the full range of formulas with the crucial steps (after the atomic variable free formulas) being $p \Vdash \exists x\psi \Leftrightarrow \exists n\,(p \Vdash \varphi(n))$; $p \Vdash \neg\varphi \Leftrightarrow \forall q \leq p(q \nVdash \varphi)$ and so $p \Vdash \forall x\varphi \Leftrightarrow \forall n \forall q \leq p\,(q \nVdash \neg\varphi(n)) \Leftrightarrow \forall n \forall q \leq p \exists r \leq q\,(r \Vdash \varphi(n))$. (The definitions for conjunction and disjunction are given by $p \Vdash \varphi \wedge \psi \Leftrightarrow p \Vdash \varphi$ and $p \Vdash \psi$ and $p \Vdash \varphi \vee \psi \Leftrightarrow p \Vdash \varphi$ or $p \Vdash \psi$.) With our restricted language we can simplify the definitions and so the calculation of the complexity of the relation $p \Vdash \varphi$.

Definition 3.22: We define the relation p *forces* φ, $p \Vdash \varphi$, by induction.

- If φ is a Σ_1 formula $\exists x_0 \phi_n(\mathtt{G}, \bar{F}, e, x_0, m_1, \ldots, m_{n-1})$ then $p \Vdash \varphi$ if and only if there is an m_0 such that $\phi_n(V(p), \bar{F}, e, m_0, m_1, \ldots, m_{n-1})$ holds (or equivalently, for every $G \supseteq V(p)$, $\phi_n(G, \bar{F}, e, m_0, m_1, \ldots, m_{n-1})$ holds.
- If φ is a Π_1 formula $\forall x_0 \neg\phi_n(\mathtt{G}, \bar{F}, e, x_0, m_1, \ldots, m_{n-1})$ then $p \Vdash \varphi$ if and only if $\forall m_0 \forall q \leq p(\neg\phi_n(V(q), \bar{F}, e, m_0, m_1, \ldots, m_{n-1})$.
- If φ is a Σ_{n+1} formula $\exists x\psi(x)$ then $p \Vdash \varphi$ if and only if $\exists m(p \Vdash \psi(m))$.
- If φ is a Π_{n+1} formula $\forall x\psi(x)$ then $p \Vdash \varphi$ if and only if $\forall m \forall q \leq p(q \nVdash \neg\psi(m)$.

Exercise 3.23: Unravel the definition for p to force a Π_2 sentence $\forall x \exists y \psi(x, y)$ to see that it means that for every m there is a n and a $q \leq p$ such that $\psi(m, n)$.

Theorem 3.24: *If $\mathcal{P}$ is a notion of forcing recursive in A then, for $n \geq 1$, forcing for Σ_n (Π_n) sentences φ (i.e. whether $p \Vdash \varphi$) is a Σ_n (Π_n) in A (and $\bar{F}$) relation.*

Proof: As we generally do from now on, we assume that the sequence $\bar{F}$ of parameters is empty and leave the relativization of results to the reader. We proceed by induction on n and for notational convenience ignore A (i.e. assume it is recursive) as well. If φ is Σ_1 or Π_1 then $p \Vdash \varphi$ is directly defined as a Σ_1 or Π_1 formula, respectively. (The point here is that the ϕ_n are uniformly recursive.) For $n \geq 1$ the result follows by induction and our definition of forcing. □

Exercise 3.25: If $p \Vdash \varphi$ and $q \leq p$ then $q \Vdash \varphi$.

We now want to tackle the question of how much genericity do we need to make forcing equal truth for generic filters/sets in the sense that if $p \Vdash \varphi$, $p \in \mathcal{G}$ and $\mathcal{G}$ is sufficiently generic then $\varphi(G)$ holds and, in the other direction, if $\varphi(G)$ holds then there is a $p \in \mathcal{G}$ such that $p \Vdash \varphi$.

Definition 3.26: The filter $\mathcal{G}$ is *n-generic* for $\mathcal{P}$ (for $n \geq 1$) if and only if for every Σ_n (in $\mathcal{P}$) subset S of $\mathcal{P}$,

$$\exists p \in \mathcal{G}(p \in S \ \vee \ \forall q \leq p(q \notin S)).$$

We say that $\mathcal{G}$ is (ω-)generic if it is n-generic for all n. Similarly the descending sequence $\langle p_n \rangle$ of conditions is *n-generic* iff for every Σ_n (in $\mathcal{P}$) subset S of $\mathcal{P}$, there is an n such that $p_n \in S$ or $\forall q \leq p_n(q \notin S)$. The sequence is called *(ω-)generic* if it is n-generic for all n. We also say that the function or set G determined by an (n-)generic filter or sequence is itself (n-)generic. These notions all relativize to an arbitrary X in the obvious way. We then say, for example, that G is *n-generic relative to (or over) X*.

The following equivalence is now immediate.

Proposition 3.27: *Let $\mathcal{C}_n$ be the class of dense sets $\{p : p \in S_e \ \vee \ \forall q \leq p(q \notin S_e)\} = D_{n,e}$ for all Σ_n (in $\mathcal{P}$) subsets S_e of $\mathcal{P}$. Then a filter $\mathcal{G}$ (or a descending sequence $\langle p_n \rangle$) is n-generic iff $\mathcal{G}$ ($\langle p_n \rangle$) is $\mathcal{C}_n$-generic.*

Exercise 3.28: If $D \subseteq \mathcal{P}$ is dense and Σ_n then D meets every n-generic $\mathcal{G}$. If D is dense below p (i.e. $\forall q \leq_{\mathcal{P}} p \exists r \leq_{\mathcal{P}} q(r \in P)$) and Σ_n then D meets every n-generic $\mathcal{G}$ containing p.

To build an n-generic $\mathcal{G}$ we proceed as in the construction of a generic for a given countable class of dense sets. We can now also calculate how hard it is to carry out this construction.

Proposition 3.29: *For any A-recursive notion of forcing $\mathcal{P}$ and each $n \geq 1$, there is an n-generic sequence $\langle p_k \rangle \leq_T A^{(n)}$ and so its associated n-generic G is also recursive in $A^{(n)}$. There is also a generic sequence $\langle p_n \rangle$ such that it and its associated G are recursive in $0^{(\omega)}$ ($\mathcal{P}^{(\omega)}$). Moreover, for any $p \in P$ we may require that $p_0 = p$ and so $V(p) \subseteq G$.*

Proof: Fix n. We build a generic sequence $\langle p_n \rangle$ for the $\mathcal{C}_n$ of Proposition 3.27 recursively in $0^{(n)}$ (as usual assuming $\mathcal{P}$ is recursive). We begin with p_0 the given $p \in P$. If we have already defined p_s we find, recursively in $0^{(n)}$, a $q \leq_{\mathcal{P}} p_s$ which is in $D_{n,s+1}$. This procedure clearly constructs the desired sequence and is recursive in $0^{(n)}$ by definition of the $D_{n,e}$. For ω-genericity one simply carries out this construction for the collection $\{D_{n,e} | n, e \in \omega)\}$ recursively in $0^{(\omega)}$. That G is recursive in $\mathcal{P}^{(n)}(\mathcal{P}^{(\omega)})$ follows from Proposition 3.17. □

Definition 3.30: We say that a condition p *decides a sentence* φ if $p \Vdash \varphi$ or $p \Vdash \neg\varphi$.

Theorem 3.31: *If $\mathcal{G}$ is n-generic and $\varphi \in \Sigma_n$ (Π_n) then there is $p \in \mathcal{G}$ which decides φ. Moreover, if $p \Vdash \varphi$ then $\varphi(G)$ holds while if $p \Vdash \neg\varphi$ then $\neg\varphi(G)$ holds. Moreover, if $q \in \mathcal{G}$ and $q \Vdash \varphi$ then $\varphi(G)$ holds.*

Proof: We proceed by induction on $n \geq 1$. Consider $\varphi = \exists x\psi(x, \mathtt{G})$. Now the set $S = \{p : p \Vdash \exists x\psi(x, \mathtt{G})\}$ is Σ_n by Theorem 3.24. So by the definition of n-genericity, either there is $p \in \mathcal{G}$ in S which forces φ or one with no extension forcing φ. If $p \in \mathcal{G}$ and $p \Vdash \exists x\psi(x, \mathtt{G})$, then (by definition) there is an n such that $p \Vdash \psi(n, \mathtt{G})$ (or $\phi_m(V(p), \ldots)$ for some m, if $n = 1$). Now by induction (or the basic properties of ϕ_m for $n = 1$), $\psi(n, G)$ holds and then so does $\exists x\psi(x, G)$ as required. On the other hand, suppose there is $p \in \mathcal{G}$ such that $(\forall q \leq p)(q \nVdash \exists x\psi(x, \mathtt{G}))$. In this case, we claim that $\neg\exists x\psi(x, G)$. If not, there would be an n such that $\psi(n, G)$ and so by induction (or definition for $n = 1$), a $q \in \mathcal{G}$ such that $q \Vdash \psi(n, \mathtt{G})$ (or $\phi_m(V(q), \ldots)$ if $n = 1$). So, $q \Vdash \exists x\psi(x, \mathtt{G})$. But, since $p, q \in \mathcal{G}$ they are compatible and there is an $r \in \mathcal{G}$ with $r \leq p, q$. This would contradict Exercise 3.25. Finally, we claim that in this case $p \Vdash \neg\varphi$. First, $(\forall q \leq p)(q \nVdash \exists x\psi(x, \mathtt{G}))$ implies that $(\forall q \leq p)(\forall x)(q \nVdash \psi(x, \mathtt{G}))$, and so $p \Vdash \forall x\neg\psi$ (by the definition of forcing) which is the same as $p \Vdash \neg\varphi$ as required. As for the last claim of the Theorem, note that there is some $p \in \mathcal{G}$ that p decides φ in the way that corresponds to the truth of $\varphi(G)$. The conditions p and q are compatible and so $p \Vdash \varphi$ and $\varphi(G)$ holds as required.

The case that φ is Π_n clearly follows as then $\neg\varphi$ is Σ_n and $\neg\neg\varphi = \varphi$.□

We now look at degree theoretic properties of sets with various amounts of genericity. We begin with some connections between genericity and lowness. The first improves Proposition 3.29. The second is specific to notions of forcing similar to that of Cohen.

Proposition 3.32: *For any notion of forcing $\mathcal{P}$ and each $n \geq 1$, there is an n-generic sequence $\langle p_k \rangle \leq_T \mathcal{P}^{(n)}$. For any n-generic G, $G^{(n)} \leq_T \mathcal{P}^{(n)} \oplus G$; and similarly for ω-generics.*

Proof: A sequence $\langle p_k \rangle$ as required exists by Proposition 3.29. Now note that the question of whether $x \in G^{(n)}$ is uniformly Σ_n and so we can find the $D_{n,e}$ that corresponds to the Σ_n formula $p \Vdash e \in G^{(n)}$ and, recursively in $\mathcal{P}^{(n)}$ ($\mathcal{P}^{(\omega)}$), find a k such that p_k forces this formula or no extension of it does. By Theorem 3.31, this determines if $x \in G^{(n)}(G^{(\omega)})$ or not. □

Exercise 3.33: If every condition in $\mathcal{P}$ has two V-incompatible extensions then we can roughly replace $\leq_T$ by $\equiv_T$ in Proposition 3.32. Indeed we can make $G^{(n)} \oplus \mathcal{P}^{(n)} \equiv_T C \equiv_T \mathcal{P}^{(n)} \oplus G \equiv_T (\mathcal{P} \oplus G)^{(n)}$. Similar results hold for the ω-jump. With Cohen forcing this gives a generalization of the Friedberg Completeness Theorem to iterations of the jump.

Proposition 3.34: *If G is n-generic for Cohen forcing then $G^{(n)} \equiv_T G \vee 0^{(n)}$. Similarly, if G is generic, $G^{(\omega)} \equiv_T G \vee 0^{(\omega)}$.*

Proof: It is immediate that for any G, $G \vee 0^{(n)} \leq_T G^{(n)}$. Thus, it suffices to show that if $\mathcal{G}$ is n-generic then $G^{(n)} \leq_T G \vee 0^{(n)}$. The formula $\varphi(e, \mathtt{G})$ which says that $e \in \mathtt{G}^{(\mathtt{n})}$ is uniformly Σ_n. Therefore, by Theorem 3.31 and the n-genericity of $\mathcal{G}$, either there is $p \in \mathcal{G}$ such that $p \Vdash \varphi(e, \mathtt{G})$ or there is $p \in \mathcal{G}$ such that $p \Vdash \neg\varphi(e, \mathtt{G})$. However, p forcing φ is a Σ_n relation and forcing $\neg\varphi$ is Π_n so to see if $e \in G^{(n)}$ we can search for a $p \in \mathcal{G}$ such that $p \Vdash \varphi(e, \mathtt{G})$ or $p \Vdash \neg\varphi(e, \mathtt{G})$. This is a $G \vee 0^{(n)}$ question since for Cohen forcing $p \in \mathcal{G}$ if and only if $V(p) \subseteq G$. By Theorem 3.31, the one forced is the true fact about G. The uniformity of this argument gives the desired result for generic G. □

Exercise 3.35: Find an A-recursive notion of forcing for which the analog of Proposition 3.34 does not hold, i.e. there is an n-generic G with $G^{(n)} \not\leq_T G \vee A^{(n)}$.

The next proposition gives almost all our previous incomparability and embeddability results in one fell swoop.

Proposition 3.36: *If $\mathcal{G}$ is Cohen 1-generic then the columns $G^{[i]} = \{\langle i, x\rangle | \langle i, x\rangle \in G\}$ of G form a very independent set, i.e. $\forall j(G^{[j]} \not\leq_T G^{[\hat{j}]})$.*

Proof: For each e we want to show that $\Phi_e^{G^{[\hat{j}]}} \neq G^{[j]}$. We consider the following set of conditions:

$$S_e = \{p : \exists x \left(\Phi_e^{p^{[\hat{j}]}}(x) \downarrow \neq p^{[j]}(x \downarrow)\right)\}.$$

Here we use the natural extension of our notation for columns of a set to finite binary strings: $p^{[j]}(\langle j, x\rangle = p(\langle j, x\rangle)$ and $p^{[j]}(\langle j, x\rangle = 0$ for $p(\langle j, x\rangle)\downarrow$ and $i \neq j$. We define $p^{[\hat{j}]}$ similarly. Since $S_e \in \Sigma_1$ and $\mathcal{G}$ is 1-generic, there is $p \in \mathcal{G} \cap S_e$ or there is $p \in \mathcal{G}$ no extension of which is in S_e. If $p \in \mathcal{G} \cap S_e$ then $p \subseteq G$ so the requirement is satisfied. Suppose that $p \subseteq G$ and $(\forall q \supseteq p) q \notin S_e$ then we claim that $\Phi_e^{G^{[\hat{\jmath}]}}$ is not total. If it were, let $\langle j, x \rangle$ be outside the domain of p. We must then have some $q \subset G$ with $q \leq p$, $q(\langle j, x \rangle) \downarrow$ and $\Phi_e^{q^{[\hat{\jmath}]}}(x) \downarrow$. Now let $\hat{q}(\langle j, x \rangle) = 1 - q(\langle j, x \rangle)$ and $\hat{q}(z) = q(z)$ for $z \neq \langle j, x \rangle$. So $\hat{q}^{[\hat{\jmath}]} = q^{[\hat{\jmath}]}$ and $\Phi_e^{q^{[\hat{\jmath}]}}(x) \downarrow = \Phi_e^{\hat{q}^{[\hat{\jmath}]}}(x) \downarrow$ but $\hat{q}(\langle j, x \rangle) \neq q(\langle j, x \rangle)$. Thus one of q and $\hat{q}$ (both of which extend p) is in S_e for the desired contradiction. □

Exercise 3.37: The Theorems and Propositions of this subsection relativize to an arbitrary X. For example, Proposition 3.36 now says that if G is 1-generic relative to X, then the independence results hold even relative to X, i.e. $\forall j (G^{[j]} \not\leq_T X \oplus G^{[\hat{\jmath}]})$.

Exercise 3.38: If G is Cohen 1-generic over X and $A, B \leq_T X$ then

$$A \leq_T B \Leftrightarrow A \oplus G \leq_T B \oplus G.$$

Also, $G \mid_T X$ if $X > 0$.

Exercise 3.39: Prove that if G is Cohen n-generic then the $G^{[i]}$ are very mutually Cohen n-generic in the sense that each $G^{[i]}$ is Cohen n-generic over $G^{[\hat{\imath}]}$.

Exercise 3.40: Translate the Exact Pair Theorem into the language of forcing. Hint: Given $\langle C_i \rangle$, define a notion of forcing $\mathcal{P}$ with conditions $\langle \alpha, \beta, n \rangle$ for $\alpha, \beta \in \omega^{<\omega}$ and $n \in \mathbb{N}$. The ordering is given by $\langle \alpha', \beta', n' \rangle \leq \langle \alpha, \beta, n \rangle$ if $\alpha' \supseteq \alpha$, $\beta' \supseteq \beta$, $n' \geq n$ and, for $i < n$, if $\alpha'(\langle i, x \rangle) \downarrow$ but $\alpha(\langle i, x \rangle) \uparrow$ then $\alpha'(\langle i, x \rangle) = C_i(x)$ and similarly for β' and β.

Exercise 3.41: Construct a 1-tree T such that every $G \in [T]$ is Cohen 1-generic. To be precise we want a function $F : \mathbb{N} \to \{0, 1, 2\}$ such that if $\{d_n\}$ lists the x such that $F(x) = 2$ in increasing order and for any $A \in 2^\omega$, we let $F_A(x) = A(n)$ if $x = d_n$ for some n and $F_A(x) = F(x)$ otherwise, then F_A is Cohen 1-generic for every $A \in 2^\omega$. Hint: make F 1-generic for conditions $p \in \{0, 1, 2\}^{<\omega}$.

Exercise 3.42: Show that the Cohen 1-generic degrees generate $\mathcal{D}$. Hint: Fix an $A \in 2^\omega$. Make the F of the previous construction 1-generic relative to A. Show that for any $j \neq k$, $(F_A^{[j]} \vee F_{\bar{A}}^{[j]}) \wedge (F_A^{[k]} \vee F_{\bar{A}}^{[k]}) \equiv_T A$ where for any i $F^{[i]}(x) = F(\langle i.x \rangle)$.

We close this subsection with a slight variation of our previous constructions that is needed in Sec. 5.4.

Proposition 3.43: *If $\mathcal{P}$ is a recursive notion of forcing and C_0 and C_1 are low sets, i.e. $C_0' \equiv_T 0' \equiv_T C_1'$ then there is a G which is 1-generic for $\mathcal{P}$ over C_0 and over C_1 so that, in particular, both $G \oplus C_0$ and $G \oplus C_1$ are low.*

Proof: Build a generic sequence meeting the dense sets $\{p : p \in S_e \vee \forall q \leq p(q \notin S_e)\} = D_{e,i}$ for all Σ_n in C_i subsets $S_{e,i}$ of $\mathcal{P}$ for $i \in \{0, 1\}$ as in the proof of Proposition 3.32. The point is that as both C_i are low, $0'$ can uniformly compute a density function for all of these sets. The argument for lowness is now exactly as above. □

Notes: Forcing in arithmetic was introduced in Feferman [1965]. It has since been used in various formulations by many people. Hinman [1969] introduced a version of n-genericity. Two important early papers applying forcing to degree theory are Jockusch [1980] in which many of the results of this subsection appear for the special but typical case of Cohen forcing and Jockusch and Posner [1978]. A systematic development of degree theory based on forcing was first presented in Lerman [1983]. Our approach attempts to both simplify and generalize previous versions. A very similar version has been presented in Cai and Shore [2012].

3.3. *Embedding lattices*

We have so far studied questions of embedding countable partial orders (and usl's) in $\mathcal{D}$ which is itself an usl. Now we know that $\mathcal{D}$ is not a lattice (Corollary 2.36) but we also know that some pairs of degrees do have greatest lower bounds in $\mathcal{D}$ (Theorem 2.29). Thus we can ask which lattices can be embedded in $\mathcal{D}$ preserving the full lattice structure. We now prove that every countable lattice can be embedded in $\mathcal{D}$.

Theorem 3.44: ***(Lattice Embedding Theorem)*** *Every countable lattice $\mathcal{L}$ with least element 0 is embeddable in $\mathcal{D}$ preserving the lattice structure and 0.*

For later convenience, we actually want to prove an *a priori* stronger statement about partial lattices.

Definition 3.45: A *partial lattice* $\mathcal{L}$ is a partial order $\leq_{\mathcal{L}}$ on its domain L together with partial functions $\wedge, \vee$ which satisfy the usual definitions when defined, i.e. if $x \wedge y = z$ then z is the greatest lower bound *(glb)* of x and y in $\leq_{\mathcal{L}}$; if $x \vee y = z$ then z is the least upper bound *(lub)* of x and y in $\leq_{\mathcal{L}}$. We say that $\mathcal{L}$ is recursive (in A) if L and $\leq_{\mathcal{L}}$ are recursive (in A) and $\vee$ and $\wedge$ are recursive (in A) functions on L.

Now, actually every partial lattice can be embedded into a lattice.

Theorem 3.46: *If $\mathcal{L}$ is a partial lattice with least element* 0 *and greatest element* 1 *then there is a lattice $\hat{\mathcal{L}}$ and an embedding $f : \mathcal{L} \to \hat{\mathcal{L}}$ which preserves* 0, 1, *order and all meets and joins that are defined in $\dot{\mathcal{L}}$.*

Proof: Consider the lattice $\mathcal{I}$ of nonempty ideals of $\mathcal{L}$, i.e. nonempty subsets I of L closed downward and under join in $\mathcal{L}$ (when defined). The ordering on $\mathcal{I}$ is given by set inclusion. Meet is set intersection and the join of I_1 and I_2 is the smallest ideal containing both of them. The map that sends $x \in \mathcal{L}$ to $I_x = \{y \in L | y \leq_{\mathcal{L}} x\}$, the principle ideal generated by x, is the desired embedding into the sublattice $\hat{\mathcal{L}}$ of $\mathcal{I}$ generated by the principle ideals. □

Thus as far as an embedding theorem is concerned, it may seem that there is no reason to use partial lattices but both effectiveness considerations and convenience come into play. It is certainly often more convenient to specify a partial lattice than to decide all the meets and joins. Thus we state our theorem for partial lattices.

Theorem 3.47: ***(Partial Lattice Embedding)*** *If $\mathcal{L}$ is a partial lattice recursive in A with least element* 0 *and greatest element* 1 *then there is an embedding $f : \mathcal{L} \to \mathcal{D}$ with $f(0) = \deg A$ which preserves order and all meets and joins that are defined in $\mathcal{L}$. Moreover, for $x \in \mathcal{L}$, $f(x)$ is uniformly recursive in $f(1)$, in the sense that we have sets G_x of degree $f(x)$ which are uniformly recursive in $f(1) \geq_T A$.*

To prove Theorem 3.47, we need some lattice theory. In particular, we use a type of lattice representations called lattice tables.

Definition 3.48: A *lattice table* for the partial lattice $\mathcal{L}$ is a collection, Θ, of maps $\alpha : L \to \mathbb{N}$ such that for every $x, y \in L$ and $\alpha, \beta \in \Theta$

(1) $\alpha(0) = 0$.
(2) If $x \leq_{\mathcal{L}} y$ and $\alpha(y) = \beta(y)$ then $\alpha(x) = \beta(x)$.

(3) If $x \not\leq_{\mathcal{L}} y$ then there are $\alpha, \beta \in \Theta$ such that $\alpha(y) = \beta(y)$ but $\alpha(x) \neq \beta(x)$.
(4) If $x \vee y = z$, $\alpha(x) = \beta(x)$ and $\alpha(y) = \beta(y)$ then $\alpha(z) = \beta(z)$.
(5) If $x \wedge y = z$ and $\alpha(z) = \beta(z)$ then there are $\gamma_1, \gamma_2, \gamma_3 \in \Theta$ such that $\alpha(x) = \gamma_1(x)$, $\gamma_1(y) = \gamma_2(y)$, $\gamma_2(x) = \gamma_3(x)$, $\gamma_3(y) = \beta(y)$. Such γ_i are called *interpolants for α and β (with respect to x, y and z).*

Notation 3.49: We define equivalence relations on Θ for each $x \in \mathcal{L}$ by $\alpha \equiv_x \beta$ if and only if $\alpha(x) = \beta(x)$. For sequences p, q from Θ of length n and $x \in L$, we say $p \equiv_x q$ if $p(k) \equiv_x q(k)$ for every $k < n$. In general, we say an equivalence relation E on a set S is *larger* or *coarser* than another one $\hat{E}$ if for every $(\forall a, b \in S)(a \equiv_{\hat{E}} b \Rightarrow a \equiv_E b)$. Similarly, E is *finer* or *smaller* than $\hat{E}$ if $(\forall a, b \in S)(a \equiv_E b \Rightarrow a \equiv_{\hat{E}} b)$. With this ordering on equivalence relations, the lub of E and $\hat{E}$ is simply their intersection. Their glb is the smallest equivalence class on S that contains their union. This is also the transitive closure of their union under the two relations.

The conditions of Definition 3.48 can now be restated in terms of these equivalence relations:

(1) $\alpha \equiv_0 \beta$ for all α and β and so $\equiv_0$ is the coarsest congruence class, i.e. the one identifying all elements.
(2) If $x \leq y$ then $\alpha \equiv_y \beta$ implies $\alpha \equiv_x \beta$ for all α and β and so $\equiv_x$ is larger than $\equiv_y$.
(3) If $x \not\leq_{\mathcal{L}} y$ then there are α and β such that $\alpha \equiv_y \beta$ but $\alpha \not\equiv_x \beta$ and so $\equiv_x$ is not larger than $\equiv_y$.
(4) If $x \vee y = z$ and $\alpha \equiv_x \beta$ and $\alpha \equiv_y \beta$ then $\alpha \equiv_z \beta$ and so $\equiv_z$ is the glb of $\equiv_x$ and $\equiv_y$.
(5) If $x \wedge y = z$ then there are $\gamma_1, \gamma_2, \gamma_3 \in \Theta$ such that $\alpha \equiv_x \gamma_1 \equiv_y \gamma_2 \equiv_x \gamma_3 \equiv_y \beta$. So $\equiv_z$ is certainly contained in the lub of $\equiv_x$ and $\equiv_y$. It is part of the theorem that we can arrange it so that chains of length three suffice to generate the entire transitive closure.

Thus a lattice table Θ produces a representation by equivalence relations with the dual ordering. A reason for reversing the order is that $\mathcal{D}$ is only an uppersemilattice. So joins always exist and we want them to correspond to the simple operation on equivalence relations of intersection. On the other hand, meets do not always exist and they then correspond to lub on equivalence relations which requires work to construct. Note that $\alpha(1)$ uniquely determines each $\alpha \in \Theta$, i.e. $\equiv_1$ is the finest congruence, i.e. equality which makes all elements distinct.

We now prove our representation theorem in terms of lattice tables.

Theorem 3.50: ***(Representation Theorem)*** *If $\mathcal{L}$ is a recursive (in A) partial lattice with 0, 1 then there is a uniformly recursive (in A) lattice table Θ for $\mathcal{L}$.*

Proof: Define $\beta_{x,i}$ for $x, y \in L$, $i = 0, 1$ by

$$\beta_{x,0}(y) = \begin{cases} \langle x, 0\rangle \text{ if } y \neq 0 \\ 0 \text{ if } y = 0 \end{cases} \qquad \beta_{x,1}(y) = \begin{cases} \beta_{x,0}(y) \text{ if } y \leq_{\mathcal{L}} x \\ \langle x, 1\rangle \text{ if } y \nleq_{\mathcal{L}} x. \end{cases}$$

The set of these $\beta_{x,i}$ satisfy (1), (2), (3) and (4). We now want to sequentially close off under adding interpolants as required in (5) for each relevant instance . To do so, we have some dovetailing procedure which does the following. Consider $x \wedge y = z$ and $\alpha \equiv_z \beta$. We want to add $\gamma_1, \gamma_2, \gamma_3$ as required in (5) and preserve the truth of (1)-(4) in the expanded set. If $x \leq_{\mathcal{L}} y$ or $y \leq_{\mathcal{L}} x$, it is easy to do so just using α and β. If not (i.e. $x \nleq_{\mathcal{L}} y$ and $y \nleq_{\mathcal{L}} x$), then choose new numbers a, b, c, d not used yet and for $w \in L$ let

$$\gamma_1(w) = \begin{cases} \alpha(s) \text{ if } w \leq_{\mathcal{L}} x \\ a \text{ if } w \nleq_{\mathcal{L}} x \end{cases} \qquad \gamma_2(w) = \begin{cases} \gamma_1(w) \text{ if } w \leq_{\mathcal{L}} y \\ b \text{ if } w \leq_{\mathcal{L}} x \text{ and } w \nleq_{\mathcal{L}} y \\ c \text{ otherwise} \end{cases}$$

$$\gamma_3(w) = \begin{cases} \beta(w) \text{ if } w \leq_{\mathcal{L}} y \\ a \text{ if } w \leq_{\mathcal{L}} x \text{ and } w \nleq_{\mathcal{L}} y \\ d \text{ otherwise.} \end{cases}$$

This is a recursive (in A) procedure and it is an Exercise to check that it works. □

Exercise 3.51: The construction given above provides a lattice table for $\mathcal{L}$.

Now we can turn to the proof of our embedding theorem for partial latices.

Proof: (**of Theorem 3.47**) We begin with a lattice table Θ for $\mathcal{L}$ which is recursive in $\mathcal{L}$. We define a notion of forcing $\mathcal{P}$ with elements $p \in \Theta^{<\omega}$, the natural ordering $p \leq_{\mathcal{P}} q$ if $p \supseteq q$ and $V(p) = p$. Our generics are then maps $G : \mathbb{N} \rightarrow L$. Define, for $x \in L$, $G_x : \mathbb{N} \rightarrow \mathbb{N}$ by $G_x(n) = G(n)(x)$. The desired embedding is given by $0 \mapsto \mathbf{a}$. For $x \neq 0$, $x \mapsto \deg(G_x) \vee \mathbf{a}$. We use a sufficient amount of genericity to prove that this map really is an embedding that preserves all the required structure. For notational convenience we assume that A is recursive but at times point out the notational changes needed when it is not. We follow the numbering of clauses in Definition 3.48.

(1) By definition, 0 is preserved by our embedding (or sent to **a** if so desired). (Note, however, that $G_0(n) = 0$ for all n and so G_0 is recursive for any $\mathcal{L}$.)
(2) Suppose $x \leq_{\mathcal{L}} y$. We must show that $G_x \leq_T G_y$. Given n, we want to compute $G_x(n) = G(n)(x)$. Find any $\alpha \in \Theta$ such that $\alpha(y) = G(n)(y) = G_y(n)$, i.e. $\alpha \equiv_y G(n)$. One exists because $G(n)$ is one such. As Θ is uniformly recursive we can search for one. Then since $x \leq_{\mathcal{L}} y$ and $G(n) \equiv_y \alpha$, by Definition 3.48(2) we have that $G(n) \equiv_x \alpha$ so $G(n)(x) = \alpha(x) = G_x(n)$.
(4) Suppose $x \vee y = z$. We must show that $G_z \equiv_T G_x \oplus G_y$. By the preservation of order, $G_z \geq_T G_x \oplus G_y$, so it suffices to compute $G_z(n) = G(n)(z)$ from $G_x(n)$ and $G_y(n)$. We search for an $\alpha \in \Theta$ such that $\alpha(x) = G(n)(x)$ and $\alpha(y) = G(n)(y)$, i.e. $\alpha \equiv_{x,y} G(n)$. There is one and we can find it as above. Now as $\alpha \equiv_{x,y} G(n)$, $\alpha \equiv_z G(n)$ by Definition 3.48(3), so $\alpha(z) = G(n)(z)$.

We can also say something about the image of 1 under the embedding. Given n, $G_1(n) = G(n)(1)$ so $G_1 \equiv_T G$ since by Definition 3.48(2) for any $\alpha \in \Theta$, $\alpha(1)$ determines α uniquely and uniformly recursively. Thus the greatest degree in the embedding is the degree of the generic G ($G \oplus A$ when $\mathcal{L}$ is not recursive).

Until this point, we have not used any genericity. We now turn to nonorder and infimum.

(3) Suppose $x \not\leq y$. We want to prove that $\Phi_e^{G_y} \neq G_x$ for every e. Suppose that G is 1-generic (over A) and consider the sets

$$S_e = \{p \in \Theta^{<\omega} : (\exists n)[\Phi_e^{p_y}(n) \downarrow \neq p_x(n)]\}$$

where $p_x \in \omega^{<\omega}$ is defined in the obvious way by $p_x(m) = p(m)(x)$. $S_e \in \Sigma_1$ because given σ we can compute $p(n)(x)$ (since Θ is uniformly recursive). Therefore, the 1-genericity of G implies that there is a $p \in \mathcal{G} \cap S_e$ or there is a $p \in \mathcal{G}$ no extension of which is in S_e. Suppose $p \in \mathcal{G} \cap S_e$, then $\Phi_e^{G_y}(n) \neq G_x(n)$ as $p_y \subset G_y$ and $p_x \subset G_x$ and we are done. Otherwise, no extension of p is in S_e. Suppose then, for the sake of a contradiction, that $\Phi_e^{G_y} = G_x$. Let α and β be as in

Definition 3.48(3) for x and y. By the obvious density of the sets $D_n = \{p | \exists m > n(p(m) = \alpha\}$ and the 1-genericity of $\mathcal{G}$, there is a $q \leq p$ and an $m > |p|$ such that $q(m) = \alpha$ and $q \in \mathcal{G}$. Moreover as $\Phi_e^{G_y}(m) \downarrow$ by our assumptions, we may also guarantee that $\Phi_e^{q_y}(m) \downarrow$ by simply choosing q as a long enough initial segment of G. Consider now the condition $\hat{q}$ such that $\hat{q}(k) = q(k)$ for $k \neq m$ and $\hat{q}(m) = \beta$. Our choice of α, β and q guarantees that $\hat{q} \leq p$, $q \equiv_y \hat{q}$ and $q \not\equiv_x \hat{q}$. Thus $\Phi_e^{q_y}(m) \downarrow = \Phi_e^{\hat{q}_y}(m) \downarrow$ but $q_x(m) \neq \hat{q}_x(m)$. So one of q and $\hat{q}$ is in S_e by definition for the desired contradiction.

(5) Suppose that $x \wedge y = z$ and $\Phi_e^{G_x} = \Phi_e^{G_y} = D$. We want to prove that $D \leq_T G_z$. Now the assertion that $\Phi_e^{G_x}$ and $\Phi_e^{G_y}$ are total and equal is Π_2. So let us assume that $\mathcal{G}$ is 2-generic (over A) and so there is (by Theorem 3.31) a $p \in \mathcal{G}$ such that p forces this sentence. Thus for each n and $q \leq p$, there is an $r \leq q$ such that $r \Vdash \Phi_e^{G_x}(n) \downarrow = \Phi_e^{G_y}(n) \downarrow$. We now wish to compute $D(n)$ from G_z. As above, we can recursively in G_z find a $q \leq p$ such that $q \Vdash \Phi_e^{G_x}(n) \downarrow = \Phi_e^{G_y}(n) \downarrow$ and $q_z \subset G_z$ (since some initial segment of G does this). We claim that $\Phi_e^{q_x}(n) = D(n)$. To see this consider a $t \in \mathcal{G}$ such that $t \leq p$, $t_z \subseteq G_z$ and $t \Vdash \Phi_e^{G_x}(n) \downarrow = \Phi_e^{G_y}(n) \downarrow$. Necessarily, $\Phi_e^{t_x}(n) \downarrow = \Phi_e^{t_y}(n) \downarrow = D(n)$ and $t \equiv_z q$. By suitably lengthening t or q we may assume that they have the same length m. Let $l = |p| < m$. We now use both the interpolants guaranteed by Definition 3.48(5) and the fact that p forces $\Phi_e^{G_x}$ and $\Phi_e^{G_y}$ to be total and equal.
For each k with $l \leq k < m$ we choose interpolants $\gamma_{k,i}$ (for $i \in \{1, 2, 3\}$) between $q(k)$ and $t(k)$ as in Definition 3.48(5). We let $q_i(k) = p(k) = t(k)$ for $k < l$ and $q_i(k) = \gamma_{k,i}$ for $l \leq k < m$. We also let $q_0 = q$ and $q_4 = t$. So $q = q_0 \equiv_x q_1 \equiv_y q_2 \equiv_x q_3 \equiv_y q_4 = t$. We now extend the q_i in turn to make them force convergence at n but remain congruent modulo z. In fact, we make a single extension for all of them. By the fact that $p \Vdash \Phi_e^{G_x} = \Phi_e^{G_y}$ and $q_1 \leq p$, we can find an $r_1 = q_1\hat{\ }s_1$ such that $r_1 \Vdash \Phi_e^{G_x}(n) \downarrow = \Phi_e^{G_y}(n) \downarrow$. We now extend $q_2\hat{\ }s_1$ to $r_2 = q_2\hat{\ }s_1\hat{\ }s_2$ such that $r_2 \Vdash \Phi_e^{G_x}(n) \downarrow = \Phi_e^{G_y}(n) \downarrow$.
Finally we extend $q_3\hat{\ }s_1\hat{\ }s_2$ to $r_3 = q_3\hat{\ }s_1\hat{\ }s_2\hat{\ }s_3$. Let $s = s_1\hat{\ }s_2\hat{\ }s_3$ and consider $q_i\hat{\ }s$ for $i \leq 4$. Looking at each successive pair we see by the alternating (between x and y) congruences that they all force the same equal values for $\Phi_e^{G_x}(n)$ and $\Phi_e^{G_y}(n)$. Thus, by transitivity of equality and permanence of computations under extension, $\Phi_e^{q_x}(n) = \Phi^{t_x}(n) = D(n)$ as required. □

By Theorem 3.24, the embedding of $\mathcal{L}$ given by the generic G produced in Theorem 3.47 can be taken to be into the degrees below the double jump of $\mathcal{L}$. We can improve this by a direct construction recursive in $0'$.

Exercise 3.52: If $\mathcal{L}$ is a recursive lattice with 0 and 1 then it can be embedded in $\mathcal{D}(\leq \mathbf{0}')$ preserving 0. Moreover, we may take the image of 1 to be low and the image of $\mathcal{L}$ to be uniformly recursive in it. This result relativizes to an arbitrary $\mathcal{L}$ and $(\deg \mathcal{L})'$. Hint: Do a direct construction of the sort done in Sec. 2 following the proof above but when it relies on 2-genericity to guarantee the existence of extensions forcing some convergence ask $0'$ instead if they exist and if not terminate the search and declare the requirement satisfied (by nonconvergence).

Alternatively we may use the following Exercise.

Exercise 3.53: The proof given above that infima are preserved used 2-genericity. Give a proof that uses only 1-genericity. Indeed, given a partial recursive lattice $\mathcal{L}$ and any 1-generic G for the recursive notion of forcing $\mathcal{P}$ of the proof of Theorem 3.47 the map from $\mathcal{L}$ to the degrees below that of G given by $x \longmapsto \deg(G_x)$ is a lattice embedding. This implies the results of the previous exercise. More specifically, G is low. Hint: Suppose that $x \wedge y = z$ and $\Phi_e^{G_x} = \Phi_e^{G_y} = D$. Consider the Σ_1 sets $T_e = \{t | \exists n(\Phi_e^{t_x}(n) \downarrow \neq \Phi_e^{t_y}(n) \downarrow\}$ and $S_e = \{s : \exists n, \exists q, s_0, s_2, r (\text{of the same length}) \; \Phi_e^{q_x}(n) \downarrow = \Phi_e^{q_y}(n) \downarrow \neq \Phi_e^{r_x}(n) \downarrow = \Phi_e^{r_y}(n) \downarrow$ and $q \equiv_x s_0 \equiv_y s \equiv_x s_2 \equiv_y r$ so $q \equiv_z r\}$ restricted to the conditions extending a t witnessing the 1-genericity condition for T_e. This also supplies a proof for Exercise 3.52.

Exercise 3.54: If $\mathcal{L}$ is a recursive lattice with 0 and 1 then it can be embedded into $\mathcal{D}(\leq \mathbf{g})$ preserving both 0 and 1 for any Cohen 1-generic $\mathbf{g}$. Hint: Show that for any infinite recursive set Θ, the degrees which are 1-generic for $\Theta^{<\omega}$ are the same as the Cohen 1-generic degrees by defining a recursive isomorphism between Θ^ω and the elements of 2^ω with infinitely many values equal to 1 that "preserves denseness".

Next, we disprove the homogeneity conjecture for $\mathcal{D}' = \langle \mathcal{D}, \leq_T, ' \rangle$. This conjecture, like the analogous one for $\mathcal{D}$, was based on the empirical fact that every theorem about the degrees or the degrees with the jump operator relativizes and so if true in $\mathcal{D}$ (or $\mathcal{D}'$) then it is true in $\mathcal{D}(\geq \mathbf{c})$ or $\mathcal{D}'(\geq \mathbf{c})$ for every $\mathbf{c}$. The conjectures asserted then that $\mathcal{D} \cong \mathcal{D}(\geq \mathbf{c})$ and even that $\mathcal{D}' \cong \mathcal{D}'(\geq \mathbf{c})$ for every degree $\mathbf{c}$.

Theorem 3.55: *There is c such that* $(\mathcal{D}, \leq, ') \not\cong (\mathcal{D}(\geq \mathbf{c}), \leq, ')$.

Proof: If not, then $[\mathbf{0}, \mathbf{0}''] \cong [\mathbf{c}, \mathbf{c}'']$ for every $\mathbf{c}$. To find a contradiction, it is sufficient (by Theorem 3.47) to find partial lattice recursive in $\mathbf{c}$ which cannot be embedded in $[\mathbf{0}, \mathbf{0}'']$.

Now it is a fact of lattice theory that there are continuum many finitely generated lattices indeed ones with only four generators. We supply ones with seven generators in the next subsection. On the other hand, only countably many finitely generated lattices can be embedded in $[0, 0'']$ since the lattice embedded is determined by the image of its generators. Thus we may choose an $\mathcal{L}$ which is finitely generated but not embeddable in $[\mathbf{0}, \mathbf{0}'']$. $\mathcal{L}$ has some degree, say $\mathbf{c}$. By theorem, $\mathcal{L}$ is embeddable in $[\mathbf{c}, \mathbf{c}'']$. Thus $[\mathbf{0}, \mathbf{0}''] \not\cong [\mathbf{c}, \mathbf{c}'']$ as required. □

Corollary 3.56: *The homogeneity conjecture for $\mathcal{D}'$ fails.*

Notes: Representations by equivalence relations is an old subject in lattice theory. In degree theory they were first used to embed all finite lattices in $\mathcal{D}$ and certain special lattices as initial segments of $\mathcal{D}$ by Thomason [1970]. The version used here in terms of tables is particularly suited to degree theory and was introduced in Lerman [1971] and extensively presented in his [1983]. Their use to embed lattices not as initial segments appears in Shore [1982] where it is used to prove Theorem 3.44 and Exercise 3.52 and various strengthenings of Theorem 3.55. The first proof of Theorem 3.55 and so the failure of the homogeneity conjecture for $\mathcal{D}'$ is due to Feiner [1970] but it depended on the construction of Σ_1 but not recursively presented Boolean algebras and known, but much more complicated, embeddings of lattices as initial segments of $\mathcal{D}$. Exercises 3.53 and 3.54 and some aspects of our treatment of lattice tables come from Greenberg and Montalbán [2003].

3.4. *Effective successor structures*

For later applications, we would like to have a specific family of size $2^{\aleph_0}$ of finitely generated partial lattices that code arbitrary sets S in a relatively simple way and can be embedded below various degrees related to S in ways that we specify later. These partial lattices begin with ones that are effective successor structures.

Definition 3.57: An *effective successor structure* is a partial lattice generated by five elements e_0, e_1, d_0, f_0, f_1 with (for each $n \geq 0$) relations

$$(d_{2n} \vee e_0) \wedge f_1 = d_{2n+1} \qquad (d_{2n+1} \vee e_1) \wedge f_0 = d_{2n+2},$$

where the d_n are all distinct (and pairwise incomparable). For any fixed $S \subseteq \omega$, we define the class of *effective successor structures* $\mathcal{L}_S$, by adding on two additional generators g_0 and g_1 and the additional relations

$$n \in S \Leftrightarrow d_n \leq g_0, g_1.$$

It is clear that the class of effective successor structures $\mathcal{L}_S$ provides us with continuum many different finitely generated partial lattices (at least one for each $S \subseteq \omega$) that we can use in the proof of Theorem 3.55.

Thus, we have represented S in a partial lattice $\mathcal{L}_S$. For later applications we now analyze the relations between the complexities of S and $\mathcal{L}_S$ or more precisely its embeddings in $\mathcal{D}$. To make these relations as simple as possible we want to impose some additional conditions on our partial lattices and slightly modify the coding procedure for S.

Definition 3.58: A *nice effective successor structure* is a partial lattice extension of an effective successor structure gotten by adding a least element 0 and additional elements b_0, b_1 and $\hat{d}_n$ for each $n \in \omega$ such that $b_0 \not\geq b_1$ and $(\forall n \in \omega)(d_n \vee b_0 \geq b_1 \;\&\; d_n \wedge \hat{d}_n = 0 \;\&\; (\forall m \neq n)(\hat{d}_n \geq d_m)$.

Note that any embedding f of a nice effective successor structure in $\mathcal{D}$ makes the images $f(d_n)$ of d_n independent and so it can be extended to an $\mathcal{L}_S$ representing S as above for any S by adding on an exact pair for the ideal generated by the $f(d_n)$ for $n \in S$. We now want to analyze the complexity of sets coded by a slightly different method in such substructures of $\mathcal{D}$.

Proposition 3.59: *If* $\mathbf{b}_0, \mathbf{b}_1, \mathbf{e}_0, \mathbf{e}_1, \mathbf{d}_0, \mathbf{f}_0, \mathbf{f}_1 \leq \mathbf{a}$ *are the generators in a nice effective successor structure (necessarily contained in* $\mathcal{D}(\leq \mathbf{a})$*) and* $\mathbf{g}_0, \mathbf{g}_1 \leq \mathbf{a}$ *then we say that* $\mathbf{g}_0, \mathbf{g}_1$ *code the set* $\hat{S} = \{n | \exists x(x \vee b_0 \geq b_1 \;\&\; x \leq \mathbf{d}_n, \mathbf{g}_0, \mathbf{g}_1\}$. *In this situation,* $\hat{S} \in \Sigma_3^A$.

Proof: We first compute the complexity of the structure $\mathcal{D}(\leq \mathbf{a})$. We represent this structure in terms of indices i such that Φ_i^A is total. (So this assigns countably many indices to each degree.) This set is Π_2^A. The order of Turing reducibility on these indices is given by $k \leq_T i$ if an only if

$$\exists j(\Phi_j^{\Phi_i^A} = \Phi_k^A) \;\Leftrightarrow\; \exists j \forall n \exists s(\Phi_{j,s}^{\Phi_{i,s}^A}(n) = \Phi_{k,s}^A(n))$$

and so is Σ_3^A. (Thus the relation that i and k represent the same degree is also Σ_3^A. We can now choose a unique representative from the class of indices coding a single set Φ_i^A uniformly in a Σ_3^A way by taking the j such that $\langle i, j\rangle$ is the first enumerated by A'' in a fixed enumeration of the pairs such that $\Phi_i^A \equiv_T \Phi_j^A$.)

Next note that there is a recursive function h on indices such that $\Phi_i^A \oplus \Phi_j^A = \Phi_{h(i,j)}^A$. So the function corresponding to join is recursive on the indices. Now infimum would naturally be Π_4 on the indices but we have added enough additional structure so as to be able of avoid using infima directly in the recovery of $\hat{S}$.

By recursion on n, we define positive Σ_1 formulas φ_n in $\leq, \vee$ (i.e. no negation symbols are used in the formula which has $\leq$ and $\vee$ but not $\wedge$ in it) such that $\mathcal{D}$ or equivalently $\mathcal{D}(\leq\mathbf{a})$ satisfies $\varphi_n(\mathbf{x})$ if and only if $\mathbf{0} < \mathbf{x} \leq \mathbf{d}_n$.

$$\varphi_0(x) \equiv x \leq \mathbf{d}_0 \ \&\ x \vee \mathbf{b_0} \geq \mathbf{b_1};$$
$$\varphi_{2n+1}(x) \equiv x \vee \mathbf{b_0} \geq \mathbf{b_1} \ \&\ \exists y(\varphi_{2n}(y) \ \&\ x \leq (y \vee \mathbf{e}_0), \mathbf{f}_1);$$
$$\varphi_{2n+2}(x) \equiv x \vee \mathbf{b_0} \geq \mathbf{b_1} \ \&\ \exists y(\varphi_{2n+1}(y) \ \&\ x \leq (y \vee \mathbf{e}_1), \mathbf{f}_0)\ .$$

It is easy to see by induction that, for any degree $\mathbf{x}$, $\varphi_n(\mathbf{x})$ is true in $\mathcal{D}$ or equivalently in $\mathcal{D}(\leq\mathbf{a})$ only if $\mathbf{0} < \mathbf{x} \leq_{\mathbf{T}} \mathbf{d}_n$ (Exercise). By our analysis of the complexity of the structure $\mathcal{D}(\leq\mathbf{a})$, the φ_n are uniformly Σ_3^A on the indices.

Note now that $n \in \hat{S}$ if and only if there is an index i such that $\Phi_i^A \oplus C \geq B$, $\varphi_n(\Phi_i^A)$ and $\Phi_i^A \leq_T G_0, G_1$ where we are using G_0, G_1, C and B for some fixed $\Phi_{g_0}^A, \Phi_{g_1}^A, \Phi_{b_0}^A$ and $\Phi_{b_1}^A$ of degrees $\mathbf{g}_0, \mathbf{g}_1, \mathbf{b}_0$ and $\mathbf{b}_1$, respectively. By the uniformity of the φ_n being Σ_3^A, this suffices to show that $\hat{S}$ is Σ_3^A. □

Remark 3.60: If $\mathbf{g}_0, \mathbf{g}_1 \leq \mathbf{a}$ are an exact pair for the ideal generated by $\{\mathbf{d}_n | n \in S\}$ then the set they code is $\{n | \mathbf{d}_n \leq \mathbf{g}_0, \mathbf{g}_1\}$ (Exercise). Thus if we can show for some embedding of a nice effective successor structure below $\mathbf{a}$ that all Σ_3^A sets are coded by an exact pair below $\mathbf{a}$ then we know that the sets coded in this structure by pairs below $\mathbf{a}$ are precisely those which are Σ_3^A.

Notes: The conditions on (nice) effective successor structures and their use in coding arithmetic come from Shore [1981] as does Proposition 3.59.

4. The Theories of $\mathcal{D}$ and $\mathcal{D}(\leq \mathbf{0}')$

In the previous sections, we talked about embeddability issues. We need to consider more in order to understand the theory of the degrees. We now approach theorems which say that the theories of (i.e. the sets of sentences true in) $\mathcal{D}$ and $\mathcal{D}(\leq 0')$ are as complicated as possible. More precisely they are of the same Turing (even $1-1$) degree as true second and first order arithmetic, respectively. The method used is interpreting arithmetic in the degree structures.

4.1. *Interpreting arithmetic*

We say that we can interpret (true first order) arithmetic in a structure $\mathcal{S}$ with parameters $\bar{p}$ if there are formulas $\varphi_D(x)$, $\varphi_+(x,y,z)$, $\varphi_\times(x,y,z)$, $\varphi_<(x,y)$ all with parameters $\bar{p}$ and one $\varphi_c(\bar{p})$ such that for any $\bar{p} \in \mathcal{S}$ such that $\mathcal{S} \vDash \varphi_c(\bar{p})$ the structure $\mathcal{M}(\bar{p})$ with domain $D(\bar{p}) = \{x \in \mathcal{S} | \mathcal{S} \vDash \varphi_D(x)\}$ and relations $+, \times$ and $<$ defined by $\varphi_+(x,y,z)$, $\varphi_\times(x,y,z)$, $\varphi_<(x,y)$, respectively, is isomorphic to true arithmetic, i.e. the natural numbers $\mathbb{N}$ with relations given by $+$, $\times$ and $<$ respectively and there is at least one such $\bar{p}$. (We are writing the operations $+$ and $\times$ in relational form $+(x,y,z) \Leftrightarrow x+y=z$ and similarly for $\times$.) In this situation, the theory of true first order arithmetic, $Th(\mathbb{N})$, i.e. the set of sentences of arithmetic in this language true in $\mathbb{N}$, is reducible to $Th(\mathcal{S})$, the set of sentences in the language of $\mathcal{S}$ true in S. Indeed, the reduction is a $1-1$ reduction. More precisely there is a recursive function T taking sentences φ of arithmetic to ones φ^T of $\mathcal{S}$ such that $\mathbb{N} \vDash \varphi \Leftrightarrow \mathcal{S} \vDash \forall \bar{p}(\varphi_c(\bar{p}) \rightarrow \varphi^T)$. The definition of T is given by induction. Atomic formulas $+(x,y,z)$, $\times(x,y,z)$ and $x<y$ are taken to $\varphi_+(x,y,z)$, $\varphi_\times(x,y,z)$, $\varphi_<(x,y)$, respectively. A formula of the form $\exists w \psi$ is taken to $\exists w(\varphi_D(w)\ \&\ \psi^T)$ while $\forall w \psi$ is taken to $\forall w(\varphi_D(w) \rightarrow \psi^T)$. It should be clear (and, if not, routine to prove) by induction that if $\mathcal{M}(\bar{p}) \cong \mathbb{N}$ then, any sentence φ (of the relational formulation of arithmetic) is true in $\mathbb{N}$ if and only if φ^T is true in $\mathcal{M}(\bar{p})$. Thus if $\varphi_c(\bar{p})$ guarantees that $\mathcal{M}(\bar{p}) \cong \mathbb{N}$, we have the desired recursive reduction from $Th(\mathbb{N})$ to $Th(\mathcal{S})$.

A second order structure is a two sorted structure (i.e. one with two sorts of variables say x and X in its language and two domains U and $W \subseteq 2^U$ over which the two types of variable range, respectively. This provides the semantics for the quantifiers $\exists x, \forall x, \exists X$, and $\forall X$ in the obvious way). The language also has relation symbols and relations on the first sort as in a standard first order language and structure. In addition, it has one relation $x \in X$ between elements of the first sort and ones of the second sort that is interpreted by true membership. We say that it is a true second order structure if $W = 2^U$, i.e. the second order quantifiers range over all subsets of the domain U of the usual first order structure. It is a model of true second order of arithmetic if the first order language is that of arithmetic, $U \cong \mathbb{N}$ and $W = 2^U$. (Note that as with true first order arithmetic there is, up to isomorphism, only one model of true second order of arithmetic.)

We extend our notion of an interpretation of arithmetic to second order structures by adding a formula $\varphi_S(x, \bar{y})$ which implies $\varphi_D(x)$. For each tuple of degrees $\bar{\mathbf{y}}$, we are thinking of $\varphi_S(x, \bar{\mathbf{y}})$ as defining the set of $n \in \mathbb{N}$ such that $\varphi_S(\mathbf{d}_n, \bar{\mathbf{y}})$ holds for $\mathbf{d}_n$ the degree corresponding to the nth element of the model in the ordering given by $\varphi_<$. We then translate the second order quantifiers by replacing each atomic formula $x \in X$ by $\varphi_S(x, \bar{y}_X)$, $\exists X\psi$ by $\exists\bar{y}_X\psi^T$ and $\forall X\psi$ by $\forall\bar{y}_X\psi^T$ where we are thinking of the $\bar{y}_X$ as coding the set X. If, as before, $\varphi_c(\bar{p})$ guarantees that the associated first order structure is isomorphic to $\mathbb{N}$ and, in addition, as $\bar{y}$ ranges over S^n (where n is the length of $\bar{y}$) the sets $S_{\bar{y}} = \{x | \varphi_S(x, \bar{y})\}$ range exactly over all subsets of $D(\bar{p})$ then it clear (or routine to prove) that, for any second order sentence φ of arithmetic, φ is satisfied in the true second order model of arithmetic if and only if $\mathcal{S} \vDash \varphi_c(\bar{p}) \to \varphi^T$. In this case we again have a recursive reduction: a sentence ψ of second order arithmetic is "true", i.e. satisfied in the model of true second order of arithmetic if an only if $\mathcal{S} \vDash \forall\bar{p}(\varphi_c(\bar{p}) \to \psi^T)$.

Our goals now are to prove that there are interpretations of true second order arithmetic in $\mathcal{D}$ and true first order arithmetic in $\mathcal{D}(\leq\mathbf{0}')$. The first we complete in this section. We actually show in the next subsection that we can code and quantify over all countable relations on $\mathcal{D}$ in a first order way by quantifying over elements of $\mathcal{D}$. From this result it is routine to get a coding as described here of second order arithmetic in $\mathcal{D}$. The results and analysis need for $\mathcal{D}(\leq \mathbf{0}')$ are mostly contained in this section but the proof also requires material from the next section as well. In each case, the correctness condition $\varphi_c(\bar{p})$ includes the translations (via T) of the axioms of a finite axiomatization of arithmetic such as Robinson arithmetic that is strong enough to guarantee that any model of the axioms in which the ordering $<$ on its domain is isomorphic to ω is actually isomorphic to $\mathbb{N}$. The crucial steps are then to prove that there are $\bar{p}$ such that $\mathcal{M}(\bar{p}) \cong \mathbb{N}$ and that there is a formula $\varphi_{\hat{c}}$ which guarantees that the ordering of $\mathcal{M}(\bar{p})$ (given by $\varphi_<(\bar{p})$) is isomorphic to ω.

We begin with $\mathcal{D}$ and coding countable subsets of pairwise incomparable degrees by using Slaman-Woodin forcing. We then show how to deal with arbitrary countable relations on degrees.

4.2. *Slaman-Woodin forcing and Th($\mathcal{D}$)*

Let $\mathbf{S} = \{\mathbf{c}_i | i \in \mathbb{N}\}$ be a countable set of pairwise incomparable degrees. We want to make $\mathbf{S}$ definable in $\mathcal{D}$ from three parameters $\mathbf{c}$, $\mathbf{g}_0$ and $\mathbf{g}_1$. The definition is that $\mathbf{S}$ is the set of minimal degrees $\mathbf{x} \leq \mathbf{c}$ such that $(\mathbf{x} \vee \mathbf{g}_0) \wedge (\mathbf{x} \vee \mathbf{g}_1) \neq \mathbf{x}$ in the strong sense that there is a $\mathbf{d} \leq \mathbf{x} \vee \mathbf{g}_0, \mathbf{x} \vee \mathbf{g}_1$ such that $\mathbf{d} \not\leq \mathbf{x}$.

Theorem 4.1: *For any set $S = \{C_0, C_1, \ldots\}$ of pairwise Turing incomparable subsets of $\mathbb{N}$ let $C = \oplus C_i$. There are then G_0, G_1 and D_i such that, for every $i \in \mathbb{N}$ and $j < 2$, $D_i \leq_T C_i \oplus G_j$ while $D_i \not\leq_T C_i$. Moreover, the C_i are minimal with this property among sets recursive in C in the sense that for any $X \leq_T C$ for which there is a D such that $D \leq_T X \oplus G_j$ ($j < 2$) but $D \not\leq_T X$ there is an i such that $C_i \leq_T X$. Indeed, there is a notion of forcing $\mathcal{P}$ recursive in C such that any 2-generic computes such G_0 and G_1. Thus for $\mathbf{c}_i, \mathbf{c}$ and $\mathbf{g}_0, \mathbf{g}_1$ the degrees of C_i, C, G_0 and G_1 respectively, the set $\mathbf{S} = \{\mathbf{c}_i | i \in \mathbb{N}\}$ is definable in $\mathcal{D}$ from the three parameters $\mathbf{c}$, $\mathbf{g}_0$ and $\mathbf{g}_1$.*

Proof: Without loss of generality we may assume that each C_i is recursive in any of its infinite subsets: simply replace C_i by the set of binary stings σ such that $\sigma \subset C_i$. The point of this assumption is that to compute C_i from some X it suffices to show that X can compute an infinite subset of C_i as then there is an infinite subset of this set recursive in X and so then is C_i.

We build G_i as required by forcing in such a way as to uniformly define the D_i from G_0 and C_i and such that D_i is also recursive in $G_1 \oplus C_i$ (although not uniformly). We begin with the coding scheme that says how we compute the D_i

Let $\{c_{i,0}, c_{i,1}, \ldots\}$ list C_i in increasing order. Our plan is that $D_i(n)$ should be $G_0^{[i]}(c_{i,n})$ and so the D_i are uniformly recursive in $G_0 \oplus C_i$. We call $\langle i, k \rangle$ a *coding location for* C_i if $k \in C_i$. To make sure that $D_i \leq_T G_1 \oplus C_i$ as well, we guarantee that $G_0^{[i]}(c_n) = G_1^{[i]}(c_n)$ for all but finitely many n. We now turn to our notion of forcing $\mathcal{P}$.

The forcing conditions p are triples of the form $\langle p_0, p_1, F_p \rangle$ where $p_0, p_1 \in 2^{<\omega}$, $|p_0| = |p_1|$, and F_p is a finite subset of ω. We let the length of condition p be $|p| = |p_0| = |p_1|$. Refinement is defined by

$$\begin{aligned} p \leq q \Leftrightarrow\ & p_0 \supseteq q_0, p_1 \supseteq q_1, F_p \supseteq F_q, \text{ and} \\ & \text{if } i \in F_q \text{ and } |q| < \langle i, c_{i,n} \rangle \leq |p| \text{ then } p_0(\langle i, c_{i,n} \rangle) = p_1(\langle i, c_{i,n} \rangle). \end{aligned}$$

This is a finite notion of forcing with extension recursive in C. The function V is defined in the obvious way: $V(p) = p_0 \oplus p_1$ so our generic object defined from a filter $\mathcal{G}$ is $G_0 \oplus G_1$ where $G_k = \cup\{p_k | p \in \mathcal{G}\}$. We use $\mathtt{G}_k$ in our language to mean the kth coordinate the generic object. Note that $C \leq_T \mathcal{P}$ as well (Exercise) and so n-generic for $\mathcal{P}$ means generic for all Σ_n^C sets.

Note that for any $\varphi \in \Sigma_1$, if $p \Vdash \varphi$ then $(p_0, p_1, \emptyset) \Vdash \varphi$ as $V(p) = V(\langle p_0, p_1, \emptyset \rangle)$. So if $q \leq p$ and $q \Vdash \psi$ for $\psi \in \Sigma_1$ then $(q_0, q_1, F_P) \Vdash \psi$ as well.

Suppose that $\mathcal{G}$ is 1-generic for $\mathcal{P}$. It is immediate from the definition of $\leq_{\mathcal{P}}$ and the density of the recursive (in $\mathcal{P}$) sets $\{p | i \in F_p\}$ that $G_0^{[i]}$ and $G_1^{[i]}$ differ on at most finitely many $n \in C_i$. (If $i \in F_p$ and $p \in \mathcal{G}$ then $G_0^{[i]}(m) = G_1^{[i]}(m)$ for $m \in C_i$ and $m > |p|$.) Thus $D_i \leq_T G_1 \oplus C_i$ as required.

We next show that $D_i \not\leq_T C_i$, that is $\Phi_e^{C_i} \neq D_i$ for each e. Suppose for the sake of a contradiction that $D_i = \Phi_e^{C_i}$ for some e (and so in particular $\Phi_e^{C_i}$ is total). Consider the Σ_1^C set

$$S_{i,e} = \{p : \exists m(p_0(\langle i, c_{i,m} \rangle) \neq \Phi_e^{C_i}(m))\}.$$

The $S_{i,e}$ are dense because if $p \in P$ and m is such that $\langle i, c_{i,m} \rangle > |p|$ then we can define $q \leq p$ by $F_q = F_p$ and for $|p| \leq j \leq \langle i, c_{i,m} \rangle$ put $q_0(j) = q_1(j) = 1 - \Phi_e^{C_i}(m)$. So $q \in S_{i,e}$ and $q \leq p$ as desired. Thus, there is a $p \in \mathcal{G} \cap S_{i,e}$ for which

$$D_i(m) = G_0(\langle i, c_{i,m} \rangle) = p_0(\langle i, c_{i,m} \rangle) \neq \Phi_e^{C_i}(m),$$

contradicting $D_i = \Phi_e^{C_i}$.

Now, we have to ensure the minimality of the C_i. In other words, we want to prove that if

$$\Phi_e^{X \oplus G_0} = \Phi_i^{X \oplus G_1} = D, \qquad X \leq_T C \qquad \text{and} \qquad D \not\leq_T X$$

then $C_k \leq_T X$ for some k. Consider the sentence φ that says that $\Phi_e^{X \oplus G_0}$ and $\Phi_i^{X \oplus G_1}$ are total and equal. It is Π_2 in C (because $X \leq_T C$) and true of $G = G_0 \oplus G_1$. So, if we now assume that $\mathcal{G}$ is 2-generic, there is $p \in \mathcal{G}$ such that $p \Vdash \varphi$. Suppose first that $\neg \exists n (\exists \sigma \supseteq p_0)(\exists \tau \supseteq p_0)[\Phi_e^{X \oplus \sigma}(n) \downarrow \neq \Phi_e^{X \oplus \tau}(n) \downarrow]$. Then we claim D is computable from X. To compute $D(n)$ search for any $\sigma \supseteq p_0$ such that $\Phi_e^{X \oplus \sigma}(n) \downarrow$ and output this value as the answer. There is such a $\sigma \subset G_0$ by the totality of $\Phi_e^{X \oplus G_0}$. Our assumption that there is no pair of extensions of p_0 that give two different answers implies that any such σ gives the answer $\Phi_e^{X \oplus G_0}(n) = D(n)$.

On the other hand, suppose there is such a splitting for n given by $p_0 \hat{\ } \sigma$, $p_0 \hat{\ } \tau$. By extending one of σ and τ if necessary, we may assume that

$|\sigma| = |\tau|$. We claim that $p_0{}^\frown\sigma$ and $p_0{}^\frown\tau$ differ at a coding location $\langle k, c_{k,m}\rangle$ for some $k \in F_p$. Let τ' be such that

$$\Phi_i^{X\oplus(p_1{}^\frown\tau{}^\frown\tau')}(n)\downarrow = \Phi_e^{X\oplus(p_0{}^\frown\tau{}^\frown\tau')}(n)\downarrow.$$

There must be such a τ' as $(p_0{}^\frown\tau, p_1{}^\frown\tau, F_p) \le p$ and so it has a further extension $q = (p_0{}^\frown\tau{}^\frown\rho_0, p_1{}^\frown\tau{}^\frown\rho_1, F_p)$ which forces $\Phi_e^{X\oplus G_0}(n)\downarrow = \Phi_i^{X\oplus G_1}(n)\downarrow$. Next consider $\hat{q} = (p_0{}^\frown\tau{}^\frown\rho_0, p_1{}^\frown\tau{}^\frown\rho_0, F_p) \le p$. It also has an extension $(p_0{}^\frown\tau{}^\frown\rho_0{}^\frown\mu_0, p_1{}^\frown\tau{}^\frown\rho_0{}^\frown\mu_1, F_p) \Vdash \Phi_e^{X\oplus G_0}(n)\downarrow = \Phi_i^{X\oplus G_1}(n)\downarrow$. It is now clear that $\tau' = \rho_0{}^\frown\mu_1$ has the desired property. Next, consider the condition $q = (p_0{}^\frown\sigma{}^\frown\tau', p_1{}^\frown\tau{}^\frown\tau', F_p)$. Notice that $q \not\le p$ because:

(1) $\Phi_e^{X\oplus(p_0{}^\frown\sigma)}(n) = \Phi_e^{X\oplus(p_0{}^\frown\sigma{}^\frown\tau')}(n)$ as $p_0{}^\frown\sigma{}^\frown\tau' \supseteq p_0{}^\frown\sigma$.
(2) $\Phi_i^{X\oplus(p_1{}^\frown\tau{}^\frown\tau')}(n) = \Phi_e^{X\oplus(p_0{}^\frown\tau)}(n)$ by our choice of τ', but
(3) $\Phi_e^{X\oplus(p_0{}^\frown\sigma)}(n) \ne \Phi_e^{X\oplus(p_0{}^\frown\tau)}(n)$ because $n, p_0{}^\frown\sigma, p_0{}^\frown\tau$ were chosen to be splitting.

Hence, $\Phi_e^{X\oplus(p_0{}^\frown\sigma{}^\frown\tau')}(n) \ne \Phi_i^{X\oplus(p_1{}^\frown\tau{}^\frown\tau')}(n)$ and so q does not extend p. However, $p_0{}^\frown\sigma{}^\frown\tau' \supseteq p_0$ and $p_1{}^\frown\tau{}^\frown\tau' \supseteq p_1$, so it must be that $p_0{}^\frown\sigma{}^\frown\tau'$ and $p_1{}^\frown\tau\ {}^\frown\tau'$ differ at a coding location above $|p|$. Therefore, $p_0{}^\frown\sigma$ and $p_0{}^\frown\tau$ differ at a coding location $\langle k, n\rangle$ with $k \in F_p$.

We now show that there must be such $p_0{}^\frown\sigma$ and $p_0{}^\frown\tau$ which differ at only one number (which then must be a coding location $\langle k, n\rangle$ for some $k \in F_p$). Suppose σ, τ are strings as above with $|\sigma| = |\tau| = \ell$. Let $\sigma = \gamma_0^0, \gamma_1^0, \ldots, \gamma_z^0 = \tau$ be a list of strings in $\{0,1\}^\ell$ such that $\gamma_i^0, \gamma_{i+1}^0$ differ at only one number for each i. Let β be such that $\Phi_e^{X\oplus(p_0{}^\frown\gamma_1^0{}^\frown\beta)}(n)\downarrow$ (such a β exists by the same argument as before). Set $\gamma_i^1 = \gamma_i^0{}^\frown\beta$ for each $0 \le i \le z$. Repeat this process for each $j \le z$. At step $j+1$, let β be such that $\Phi_e^{X\oplus(p_0{}^\frown\gamma_{i+1}^j{}^\frown\beta)}(n)\downarrow$, and set $\gamma_i^{j+1} = \gamma_i^j{}^\frown\beta$ for each $0 \le i \le z$. At the end, we have strings $\gamma_0^z, \gamma_1^z, \ldots, \gamma_z^z$ such that $\Phi_e^{X\oplus(p_0{}^\frown\gamma_i^z)}(n)\downarrow$ for each i, and $p_0{}^\frown\gamma_i^z, p_0{}^\frown\gamma_{i+1}^z$ differ at only one number for each i. Since

$$\Phi_e^{X\oplus(p_0{}^\frown\gamma_0^z)}(n) = \Phi_e^{X\oplus(p_0{}^\frown\sigma)}(n) \ne \Phi_e^{X\oplus(p_0{}^\frown\tau)}(n) = \Phi_e^{X\oplus(p_0{}^\frown\gamma_z^z)}(n),$$

there must be an i for which $\Phi_e^{X\oplus(p_0{}^\frown\gamma_i^z)}(n) \ne \Phi_e^{X\oplus(p_0{}^\frown\gamma_{i+1}^z)}(n)$. The strings $p_0{}^\frown\gamma_i^z, p_0{}^\frown\gamma_{i+1}^z$ differ at only one number and it must be a coding location $\langle k, m\rangle$ for some $k \in F_p$ as required.

Next, we show that X can find infinitely many coding locations $\langle k, m\rangle$ for some fixed $k \in F_p$. Suppose we want to find such a location $\langle k, m\rangle$ with $m > M$. Search for strings $p_0{}^\frown\sigma$ and $p_0{}^\frown\tau$ that agree on the first M positions, differ at only one position, and satisfy $\Phi_e^{X\oplus(p_0{}^\frown\sigma)}(n) \ne \Phi_e^{X\oplus(p_0{}^\frown\tau)}(n)$.

Such strings must exist because we could have started the above analysis at any condition $q \in \mathcal{G}$ with $q \leq p$ (so we can find such strings agreeing on arbitrarily long initial segments). The position at which $p_0\hat{\ }\sigma$ and $p_0\hat{\ }\tau$ differ must be a coding location bigger than M. Since F_p is finite, infinitely many of these coding locations must be for the same k. Given this k, X can find infinitely many coding locations $\langle k, c_{k,m}\rangle$. Hence, X can enumerate an infinite subset of C_k and so can compute C_k by our initial assumption on the C_i. □

As 2-genericity sufficed for the proof of the theorem above, we can get the required $G_j \leq_T C''$ and, indeed with $(G_0 \oplus G_1)'' \equiv_T C''$. We show below (Theorem 4.6 and Exercise 4.8) that we can do better.

Now we work toward coding arbitrary countable relations on $\mathcal{D}$.

Proposition 4.2: *If H is Cohen 1-generic over C, then, for any $i, j \in \omega$ and $X, Y \leq C$, if $X \oplus H^{[i]} \leq Y \oplus H^{[j]}$ then $i = j$ and $X \leq Y$.*

Proof: Suppose that for some e, $X, Y \leq_T C$, $\Phi_e^{Y \oplus H^{[j]}} = X \oplus H^{[i]}$ and consider the set

$$S_e = \{\sigma \in 2^{<\omega} : \exists n(\Phi_e^{Y \oplus \sigma^{[j]}}(n) \downarrow \neq X \oplus \sigma^{[i]}(n))\}.$$

$S_e \in \Sigma_1(C)$ so either there is a $\sigma \in S_e \cap H$ or there is a $\sigma \subset H$ no extension of which is in S_e. The first alternative clearly violates our assumption that $\Phi_e^{Y \oplus H^{[j]}} = X \oplus H^{[i]}$ and so there is a $\sigma \subset H$ such that $\tau \notin S_e$ for all $\tau \supseteq \sigma$. Let $n = |\sigma^{[i]}|$. If $i \neq j$ and there were $\beta \supseteq \sigma^{[j]}$ such that $\Phi_e^{Y \oplus \beta}(2n+1) \downarrow$, we could extend σ to a τ such that $\tau^{[j]} = \beta$ and $\tau^{[i]}(n) = 1 - \Phi_e^{Y \oplus \beta}(2n+1)$ (as the value of $\tau^{[i]}(n)$ is independent of $\tau^{[j]}$). In this case, we have

$$\Phi_e^{Y \oplus \tau^{[j]}}(2n+1) \downarrow \neq \tau^{[i]}(n) = (X \oplus \tau^{[i]})(2n+1)$$

and so $\tau \in S_e$, contradicting our choice of σ. Therefore, there can be no $\beta \supseteq \sigma^{[j]}$ making $\Phi_e^{Y \oplus \beta}(2n+1)$ converge while $\Phi_e^{Y \oplus H^{[j]}}$ is total by assumption and $\sigma^{[j]} \subset H^{[j]}$ for a contradiction. Thus $i = j$.

Next, we show that $X \leq_T Y$. To compute $X(n)$ from Y, search for a $\tau \supseteq \sigma$ such that $\Phi_e^{Y \oplus \tau^{[j]}}(2n)$ converges (such a τ exists because $\Phi_e^{Y \oplus H^{[i]}}$ is total and $\sigma^{[j]} \subset H^{[j]}$). Then, as usual, we claim that $\Phi_e^{Y \oplus \tau^{[j]}} = (X \oplus \tau^{[i]})(2n) = X(n)$ for if not, $\tau \in S_e$ and extends σ for a contradiction. □

Theorem 4.3: *Every countable relation $R(x_0, \ldots, x_{n-1})$ on $\mathcal{D}$ is definable from parameters. Indeed, if C is a uniform upper bound on representatives C_i of the sets with degrees $\mathbf{c}_i$ in the domain of R as well as of the*

$\langle C_{j_0},\ldots,C_{j_{n-1}}\rangle$ *such that* $R(\mathbf{c}_{j_0},\ldots,\mathbf{c}_{j_{n-1}})$ *and* H *is Cohen* 1*-generic over* C *then there is a notion of forcing recursive in* $C \oplus H$ *such that any* 2*-generic computes the required parameters. Moreover, for each* n *there is a formula* $\varphi_n(x_0,\ldots,x_{n-1},\bar{y})$ *with* $\bar{y}$ *of length some* $k > 0$ *(depending only on* n*) which includes the clauses that* $x_i \leq y_0$ *for each* $i < n$ *such that as* $\bar{\mathbf{p}}$ *ranges over all* k*-tuples of degrees, the sets of* n*-tuples of degrees* $\{\bar{\mathbf{a}}|\mathcal{D} \vDash \varphi(\bar{\mathbf{a}},\bar{\mathbf{p}})\}$ *range over all countable* n*-ary relations on* $\mathcal{D}$.

Proof: We take $\mathbf{c} = \deg(C)$ to be our first parameter. Let H be Cohen 1-generic over C and $\mathbf{h}_{i,j}$ be the degree of $H^{[\langle i,j\rangle]}$. We code R using the following countable sets of pairwise incomparable degrees.

$$\mathcal{H}_i = \{\mathbf{h}_{i,j}|j \in \mathbb{N}\} \text{ for } i < n$$

$$\mathcal{F}_i = \{\mathbf{c}_j \vee \mathbf{h}_{i,j}|j \in \mathbb{N}\} \text{ for } i < n$$

$$\mathcal{R} = \{\mathbf{h}_{0,j_0} \vee \mathbf{h}_{1,j_1} \vee \cdots \vee \mathbf{h}_{n-1,j_{n-1}} : R(\mathbf{c}_{j_0},\mathbf{c}_{j_1},\ldots,\mathbf{c}_{j_{n-1}})\}.$$

Each of these sets consists of pairwise incomparable degrees. The first and third by Proposition 3.36 that for a Cohen 1-generic H the sets $H^{[k]}$ form a very independent set. (So, for any finite sets A and B of $\mathbf{h}_{i,j}$, $\vee\{\mathbf{x}|\mathbf{x} \in A\} \leq \vee\{\mathbf{x}|\mathbf{x} \in B\}$ if and only if $A \subseteq B$.) The elements of each $\mathcal{F}_i$ are pairwise incomparable by Proposition 4.2. Our defining formula φ for R is now

$$\&_{i<n}(\mathbf{x}_i \leq \mathbf{c}) \; \& \; (\exists \mathbf{y}_i)_{i<n}(\mathbf{y}_i \in \mathcal{H}_i \; \& \; \&_{i<n}(\mathbf{x}_i \vee \mathbf{y}_i) \in \mathcal{F}_i \; \& \; \vee\{\mathbf{y}_i|i < n\} \in \mathcal{R})$$

where we understand membership in the sets $\mathcal{H}_i$, $\mathcal{F}_i$ and $\mathcal{R}$ as being defined by the appropriate formulas and parameters as given by Theorem 4.1. This also supplies the notion of forcing required in our Theorem by taking (the disjoint union of) three versions of the one provided in Theorem 4.1 for the families of pairwise Turing incomparable sets needed for these definitions as they are uniformly recursive in $C \oplus H$. The verification that this formula defines the relation is straightforward. If $R(\bar{\mathbf{x}})$ then every element of the sequence $\bar{\mathbf{x}}$ is below $\mathbf{c}$ and is therefore equal to a $\tilde{\mathbf{c}}_{j_i}$ (for $i < n$). The degrees $\mathbf{h}_{i,j_i} \in \mathcal{H}_i$ then are the witness $\mathbf{y}_i$ required in φ. In the other direction, if φ holds of any n-tuple then all its elements are below $\mathbf{c}$ and we need to consider the situation where $\varphi(\mathbf{c}_{j_0},\ldots\mathbf{c}_{j_{n-1}})$ for some j_i, $i < n$. Let the required witnesses be $\mathbf{y}_i$. As $\mathbf{y}_i \in \mathcal{H}_i$ and $(\mathbf{c}_{j_i} \vee \mathbf{y}_i) \in \mathcal{F}_i$, $\mathbf{y}_i = \mathbf{h}_{i,j_i}$. Then as $\bigvee_{i<n} \mathbf{y}_i \in \mathcal{R}$, $R(\mathbf{x}_{j_0},\mathbf{x}_{j_1},\ldots,\mathbf{x}_{j_{n-1}})$. The assertions in the Theorem about the form of the required φ are now immediate from Theorem 4.1. □

Note that with the above assumptions on $\mathbf{c}$ in this proof, Theorem 4.1, the remarks immediately following it and Proposition 3.29, we can get all the parameters need for this definition of R below $\mathbf{c}''$. We improve this by one jump in the next subsection.

We can now precisely characterize the complexity of $Th(\mathcal{D})$ as that of true second order arithmetic.

Theorem 4.4: $Th(\mathcal{D}, \leq) \equiv_1 Th^2(\mathbb{N}, \leq, +, \times, 0, 1)$.

Proof: That $Th(\mathcal{D}, \leq) \leq_1 Th^2(\mathbb{N}, \leq, +, \times, 0, 1)$ is easy. As $A \leq_T B$ is definable in arithmetic (indeed as we have seen it is Σ_3 in A and B) and quantification over all sets gives quantification over all degree, we can recursively translate any sentence about $\mathcal{D}$ to an equivalent one of about second order arithmetic. For the other direction we use the formulas φ_1, φ_2 and φ_3 of Theorem 4.3 to give an interpretation of true second order arithmetic in $\mathcal{D}$. We consider sequences of parameters $\bar{p}_D$, $\bar{p}_+$, $\bar{p}_\times$ and $\bar{p}_<$ so that $\varphi_1(\bar{p}_D)$ defines a countable set of degrees and plays the role of φ_D for our interpretation. Our correctness condition then includes the sentences that say that $\varphi_3(\bar{p}_+)$, $\varphi_3(\bar{p}_\times)$ and $\varphi_2(\bar{p}_<)$ (playing the roles of φ_+, $\varphi_\times$ and $\varphi_<$, respectively) define relations on the countable set defined by $\varphi_1(\bar{p}_D)$ to determine a structure $\mathcal{M}(\bar{p})$ (where $\bar{p}$ is the concatenation of all the sequences of parameters used here) that satisfies all the axioms of our finite theory of arithmetic. Theorem 4.3 then says that there are choices of these parameters such that the structure so defined is isomorphic to $\mathbb{N}$. After all, $\mathbb{N}$ is just a countable set with two ternary relations and one binary one. We now use $\varphi_1(x, \bar{q}) \wedge \varphi_1(x, \bar{p}_D)$ as the φ_S required for our interpretation of true second order arithmetic. Again by Theorem 4.3, as $\bar{q}$ ranges over tuples of degrees, the subsets of $M(\bar{p})$ defined by φ_s range over all subsets of $M(\bar{p})$ as required. All that remains to do is to show that we can extend the list of correctness conditions that guarantee that $\mathcal{M}(\bar{p})$ is a model of our finite axiomatization of arithmetic to also guarantee that it is isomorphic to $\mathbb{N}$. We can do this by adding on the sentence which asserts that every nonempty subset of $M(\bar{p})$ (as given by $\varphi_S(\bar{q}, \bar{p})$ for some $\bar{q}$) has an $<_\mathcal{M}$ least element, i.e. $\forall \bar{q}\{\exists x(\varphi_S(x, \bar{q}, \bar{p})) \rightarrow \exists x[\varphi_S(x, \bar{q}, \bar{p}) \wedge \neg\exists y(\varphi_S(y, \bar{q}, \bar{p}) \wedge \varphi_<(y, x, \bar{p}_<))]\}$. □

Exercise 4.5: If $\mathcal{C}$ is a jump ideal of $\mathcal{D}$ (i.e. a downward closed subset that is also closed under jump and join), then the theory of $\mathcal{C}$ is 1-1 equivalent to that of the model of second order arithmetic where set quantifiers range over the sets with degrees in $\mathcal{C}$.

Notes: Slaman and Woodin forcing was introduced in Slaman and Woodin [1986] where they proved Theorems 4.1 and 4.3. Theorem 4.4 (which as presented here follows easily from these results) is originally due to Simpson [1977] although with a very different proof using then new initial segments results and Theorem 2.33. Another version using simpler codings and previously know initial segment results along with Theorem 2.33 is in Nerode and Shore [1980]. Exercise 4.5 is from Nerode and Shore [1980a].

4.3. $Th(\mathcal{D} \leq \mathbf{0}')$

We now want to improve our coding results so that they become applicable below $\mathbf{0}'$. We begin with the Slaman and Woodin coding of sets of pairwise incomparable degrees.

Theorem 4.6: *For any set $S = \{C_0, C_1, \ldots\}$ of pairwise Turing incomparable subsets of $\mathbb{N}$ let $C = \oplus C_i$. There are then G_0,$G_1 \leq_T C'$ and D_i such that, for every $i \in \mathbb{N}$ and $j < 2$,, $D_i \leq_T C_i \oplus G_j$ while $D_i \nleq_T C_i$. Moreover, the C_i are minimal with this property among sets recursive in C in the sense that for any $X \leq_T C$ for which there is a D such that $D \leq_T X \oplus G_j$ ($j < 2$) but $D \nleq_T X$ there is an i such that $C_i \leq_T X$.*

Proof: We follow the ideas of the proof of Theorem 4.1 but replace the uses of 2-genericity for extending conditions to make something converge. At various steps we ask if there are appropriate extensions, if so we take them and continue our construction. If not we have a condition that forces some functional to diverge and so can satisfy the relevant requirement in that way. We build $D_i \leq_T G_0 \oplus C_i, G_1 \oplus C_i$ such that $D_i \nleq_T C_i$. The requirements for diagonalization here are:

$$P_{e,i} : \Phi_e^{C_i} \neq D_i.$$

Let $X_j = \Phi_j^C$. We also have requirements for minimality:

$$R_{e,,j} : \Phi_e^{G_0 \oplus X_j} = \Phi_e^{G_1 \oplus X_j} = D \Rightarrow D \leq_T X_j \text{ or } \exists i (C_i \leq_T X_j).$$

Note that we are using the same index for computing from both $X \oplus G_0$ and $X \oplus G_1$ rather than two distinct ones. This is equivalent to our previous use of two indices, say l_0 and l_1. The point is that we know that G_0 and G_1 are different. Say $G_0(x) = 0$ while $G_1(x) = 1$ for some x. Given any indices l_0 and l_1, we can find an e such that for any oracle Z, $\Phi_e^Z = \Phi_{l_0}^Z$ if $Z(x) = 0$ and $\Phi_e^Z = \Phi_{l_1}^Z$ if $Z(x) = 1$. Using this e for computing from both $X \oplus G_0$ and $X \oplus G_1$ then gives the same results as using l_0 and l_1 to compute from $X \oplus G_0$ and $X \oplus G_1$, respectively. This notational device is known as Posner's trick (or at least as one variant thereof).

We list all the requirements as Q_s. We build G_0, G_1 by finite approximations $\gamma_{0,s}, \gamma_{1,s}$ of equal length. As before we let $D_i(m) = G_0(\langle i, c_{i,m}\rangle)$ where $\{c_{i,m}\}$ is an enumeration of C_i in increasing order. So $D_i \leq_T G_0 \oplus C_i$. We guarantee that $D_i \leq_T G_1 \oplus C_i$ as before by making sure that, for each i, $G_0(\langle i, c_{i,m}\rangle \neq G_1(\langle i, c_{i,m}\rangle$ for at most finitely many m. In particular we institute a *rule for the construction* that when we act to satisfy requirement Q_n at stage s by extending the current values of γ_k $(k = 0, 1)$ we require, for $i \leq n$, $\langle i, m\rangle \geq |\gamma_{0,s}| = |\gamma_{1,s}|$ and $m \in C_i$, that the extensions γ'_k are such that $\gamma'_0(\langle i, m\rangle) = \gamma'_1(\langle i, m\rangle)$. As we act to satisfy any Q_n at most once, this rule guarantees that there are at most finitely many relevant differences between G_0 and G_1 for each i.

At stage s, if $Q_s = P_{e,i}$, we act to satisfy $P_{e,i}$. Choose m such that $\langle i, c_{i,m}\rangle \geq |\gamma_{0,s}|$. Ask if $\Phi_e^{C_i}(m)\downarrow$. If not, let $\gamma_{k,s+1} = \gamma_{k,s}$ for $k = 0, 1$. (As usual this satisfies $P_{e,i}$.) If it does converge, extend each of $\gamma_{0,s}, \gamma_{1,s}$ by the same string σ to $\gamma_{0,s+1}, \gamma_{1,s+1}$ with $\gamma_{0,s+1}(\langle i, c_{i,m}\rangle) \neq \Phi_e^{C_i}(m)$. This also satisfies the requirement because $D_i(m) = G_0(\langle i, c_{i,m}\rangle)$ by definition and trivially obeys the rule of the construction.

Note that C' can decide if $\Phi_e^{C_i}(m)\downarrow$, so this action is recursive in C'.

If $Q_s = R_{e,j}$, this stage has a substage for each requirement $Q_n = R_{e',j'}$ with $n \leq s$ that has not yet been satisfied. For notational convenience we write γ_k for $\gamma_{k,s}$ in the description of our action at stage s. At the end of each substage we define successive extensions $\gamma_{k,l}$ of γ_k satisfying the rule of the construction. We first try to satisfy $R_{e,j}$ (which, of course, we have not attempted to satisfy before). We ask if $\exists x \exists \sigma_k \supseteq \gamma_k$ which satisfy the rule of our construction and such that the $\sigma_k \oplus X$ e-split at x, i.e.

$$\Phi_e^{\sigma_0 \oplus X_j}(x)\downarrow \neq \Phi_e^{\sigma_1 \oplus X_j}(x)\downarrow .$$

Note that, when we are acting to satisfy any Q_n, checking if extensions of the current values of γ_k satisfy the rule of the construction is recursive in $\oplus\{C_i | i \leq n\}$ and so uniformly recursive in C. Thus this question can be answered by C'. There is one subtlety here. We must be careful with what we mean by a computation from X_j as there is no list of all the sets recursive in C that is uniformly recursive in C. So what we mean here is that there is a computation of Φ_j^C providing a long enough initial segment of X_j so as to make the desired computations at m converge. This makes the whole question one that is Σ_1^C and so recursive in C'.

If the answer is yes, choose as usual the first such extensions (in a uniform search recursive in C) as $\gamma_{0,0}, \gamma_{1,1}$. Note that we have now satisfied $R_{e,j}$. If the answer is no, ask if $\exists x \exists \sigma, \tau\, ((\gamma_0 \hat{}\, \sigma \oplus X_j)|_e(\gamma_0 \hat{}\, \tau \oplus X))$ (see Definition 2.30). This question is also $\Sigma_1(C)$.

- If not, let $\gamma_{k,s,0} = \gamma_{k,s}$. Then, as usual, if $\Phi_e^{G_0 \oplus X_j}$ is total, it is recursive in X as we guarantee that $G_0 \supseteq \gamma_{0,0}$. To calculate it at x, find any σ such that $\Phi_e^{\gamma_0 \hat{}\, \sigma \oplus X_j}(x) \downarrow$. This computation must give right answer. So in this case we have also satisfied $R_{e,j}$.
- If so,we can find such σ and τ (recursively in C). We interpolate between σ, τ with strings $\sigma = \delta_0 = \delta_1, \ldots, \delta_z = \tau$ which differ successively at exactly one number. Ask if $\exists \sigma_1$ such that $\Phi_e^{\gamma_0 \hat{}\, \delta_1 \hat{}\, \sigma_1 \oplus X_j}(x) \downarrow$. If not, let $\gamma_{k,0} = \gamma_k \hat{}\, \delta_1$. Note that this extension satisfies the rule of the construction and that we have satisfied $R_{e,j}$ by guaranteeing that $\Phi_e^{G_0 \oplus X_j}(x) \uparrow$. If yes, consider $\delta_2 \hat{}\, \sigma_1$ and ask again if there is a σ_2 such that $\Phi_e^{\delta_2 \hat{}\, \sigma_1 \hat{}\, \sigma_2 \oplus X_j}(x) \downarrow$. If not, let $\gamma_{k,0} = \delta_2 \hat{}\, \sigma_1$ as before obeying the rule of the construction and satisfying $R_{e,j}$. If so, we continue on inductively through the δ_k.
- Eventually we either define $\gamma_{k,0}$ and satisfy $R_{e,j}$ or we find $\sigma_1, \ldots, \sigma_z$ such that $\Phi_e^{\gamma_0 \hat{}\, \delta_l \hat{}\, \rho \oplus X_j}(x) \downarrow$ for every $l \leq z$ where $\rho = \sigma_1 \hat{}\, \ldots \hat{}\, \sigma_z$. In the second case, we set $\gamma_{k,0} = \gamma_{k,s}$. This action does not satisfy $R_{e,i}$ but it demonstrates that there are $\hat{\sigma}$ and $\hat{\tau}$ which differ at exactly one number and for which $(\gamma_0 \hat{}\, \hat{\sigma} \oplus X)|_e(\gamma_0 \hat{}\, \hat{\tau} \oplus X)$. The point here is that, as $\Phi_e^{\gamma_0 \hat{}\, \delta_0 \hat{}\, \rho \oplus X_j}(x) \downarrow = \Phi_e^{\gamma_0 \hat{}\, \sigma \oplus X_j}(x) \downarrow \neq \Phi_e^{\gamma_0 \hat{}\, \tau \oplus X_j}(x) \downarrow = \Phi_e^{\gamma_0 \hat{}\, \delta_z \hat{}\, \varepsilon \oplus X_j}(x) \downarrow$, there is an l such that $\Phi_e^{\gamma_0 \hat{}\, \delta_l \hat{}\, \rho \oplus X_j}(x) \downarrow \neq \Phi_e^{\gamma_0 \hat{}\, \delta_{l+1} \hat{}\, \rho \oplus X_j}(x) \downarrow$ while $\delta_l \hat{}\, \rho$ and $\delta_{l+1} \hat{}\, \rho$ differ at exactly one number. Now consider $\gamma_1 \hat{}\, \hat{\sigma}$. If there is no μ such that $\Phi_e^{\gamma_1 \hat{}\, \hat{\sigma} \hat{}\, \mu \oplus X_j}(x) \downarrow$ then we can again satisfy $R_{e,j}$ by setting $\gamma_{k,s,0} = \gamma_{k,s} \hat{}\, \vartheta$. If there is such a μ, we compare $\Phi_e^{\gamma_1 \hat{}\, \hat{\sigma} \hat{}\, \mu \oplus X_j}(x) \downarrow$ with $\Phi_e^{\gamma_0 \hat{}\, \hat{\sigma} \hat{}\, \mu \oplus X_j}(x) \downarrow$ and $\Phi_e^{\gamma_0 \hat{}\, \hat{\tau} \hat{}\, \mu \oplus X_j}(x) \downarrow$. As the last two are different one of them must be different from the first. If $\Phi_e^{\gamma_1 \hat{}\, \hat{\sigma} \hat{}\, \mu \oplus X_j}(x) \downarrow \neq \Phi_e^{\gamma_0 \hat{}\, \hat{\sigma} \hat{}\, \mu \oplus X_j}(x) \downarrow$, we would contradict our assumption that the answer to our very first question was no as $\gamma_1 \hat{}\, \hat{\sigma} \hat{}\, \mu$ and $\gamma_0 \hat{}\, \hat{\sigma} \hat{}\, \mu$ certainly satisfy the rule of the construction. If $\Phi_e^{\gamma_1 \hat{}\, \hat{\sigma} \hat{}\, \mu \oplus X_j}(x) \downarrow \neq \Phi_e^{\gamma_0 \hat{}\, \hat{\tau} \hat{}\, \mu \oplus X_j}(x) \downarrow$, the only way we would not have the same contradiction is if the one point at which $\hat{\sigma}$ and $\hat{\tau}$ differ is a coding location $\langle k, c_{k,m} \rangle$ with $k < s$. Thus the only way our actions at this stage do not satisfy $R_{\langle e,j \rangle}$ is if there are $\hat{\sigma} \hat{}\, \mu$ and $\hat{\tau} \hat{}\, \mu$ which differ at exactly one point such that $(\gamma_1 \hat{}\, \hat{\sigma} \hat{}\, \mu \oplus X_j)|_e(\gamma_0 \hat{}\, \hat{\tau} \hat{}\, \mu \oplus X_j)$ and for any such $\hat{\sigma}$ and $\hat{\tau}$ the point of difference must be a coding location

$\langle k, c_{k,m}\rangle$ with $k < s$.

- In this last case we set $\gamma_{0,0} = \gamma_0$ and $\gamma_{1,0} = \gamma_{1,s}$. In any event, we now proceed to extend $\gamma_{1,0}$ (and then γ_1) in the same way but attempting to satisfy each $Q_n = R_{e',j'}$ with $n < s$ that has not yet been satisfied. After some finite number of such attempts we have tried them all, satisfying some and for the others producing one more example of an x and two strings $\hat{\sigma}$ and $\hat{\tau}$ differing at one number only (after $|\gamma_0|$) such that $(\gamma_0\hat{\ }\hat{\sigma} \oplus X_{j'})|_e(\gamma_1\hat{\ }\hat{\tau} \oplus X_{j'})$ for each $\langle e', j'\rangle$ which we have not yet satisfied and a guarantee that any two such strings differ at a coding location $\langle k, c_{k,m}\rangle$ with $k < n$.
- At the end of this process we let $\gamma_{k,s+1}$ be the final extension of γ_k that we have produced.

We now claim that all the requirements are satisfied. It is immediate that $P_{e,i}$ is satisfied when we act for $Q_s = P_{e,i}$ at stage s. Consider any $R_{e,j} = Q_{s_0}$. If we ever act so as to satisfy it at some stage s of the construction, it is clearly satisfied and we never act for it again. As we violate the rule of the construction at some $\langle k, c_{k,m}\rangle$ only when we act to satisfy requirement Q_n for $n \leq k$ and we do so at most once for each n, $D_i \leq_T G_1 \oplus C_i$ as required.

Finally, suppose that the first requirement that we never act to satisfy during the construction is Q_n. It must be some $R_{e,j}$. Suppose that all requirements Q_r for $r < n$ have been satisfied by stage $s_0 > n$. At each stage $s > s_0$ with $Q_s = R_{e',j'}$ we attempt to satisfy $R_{e,j}$ at some substage of the construction. As we fail, there are $\Phi_{e'}^{\delta_0 \oplus X_{j'}}(x)\downarrow \neq \Phi_{e'}^{\delta_1 \oplus X_{j'}}(x)\downarrow$ with $\delta_k \supseteq \gamma_{k,s} \supseteq \gamma_{k,n}$ which differ at exactly one point and any such pair differ at a coding location $\langle k, c_{m,k}\rangle$ with $k \leq n$. Recursively in X_j we can then search for and find infinitely many extensions δ_k of $\gamma_{k,n}$ with this property with the points at which they differ becoming arbitrarily large (as $|\gamma_{k,s}|$ is clearly going to infinity). As there are only finitely many $k \leq n$, there must be one $k \leq n$ for which infinitely many of these δ_k differ at a point of the form $\langle k, z\rangle$ with infinitely many different z. As every such point is a coding location, recursively in X we can compute an infinite subset of C_k, so by our initial assumption that each C_i is recursive in everyone of its infinite subsets $C_k \leq_T X_j$ as required for $R_{e,j}$ to be satisfied in the end. □

This step-by-step construction is the much the same as the forcing argument above, but grittier, and we gain a quantifier. This helps us determine the true complexity of $Th(\mathcal{D}, \leq \mathbf{0}')$: $Th(\mathcal{D}, \leq \mathbf{0}') \equiv_{1-1} Th(\mathbb{N}, +, \times, \leq)$.

Exercise 4.7: It is easy to show that the G_i of Theorem 4.6 can be made to have (or already have) jumps below C'.

Exercise 4.8: With the notation as in Theorem 4.1 show that for any $\mathcal{G}$ 1-generic for $\mathcal{P}$, G_0 and G_1 have the properties required by the Theorem. So in particular, we can make $G_0' \equiv_T 0' \equiv_T G_1'$. This then supplies the analogous result for Theorem 4.3, i.e. a notion of forcing recursive in the appropriate $C \oplus H$ such that any 1-generic computes the parameters necessary to define the given relation. Hint: This is not easy. A proof can be found in Greenberg and Montalbán [2003].

Theorem 4.9: *If R is an n-ary relation on $\mathcal{D}(\leq \mathbf{0}')$ which is uniformly recursive in a low degree $\mathbf{c}$ in the sense that there are families of sets $\{X_i\} = S$ and $\{\langle X_{i_1}, \ldots, X_{i_n}\rangle\} = T$ uniformly recursive in $C \in \mathbf{c}$ such that $\{\deg(X_i) | X_i \in S\}$ is the field of R (i.e. all elements that occur in any n-tuple satisfying R) and $\{\langle \deg X_{i_1}, \ldots, \deg X_{i_n}\rangle \mid \langle X_{i_1}, \ldots, X_{i_n}\rangle \in T\} = R$, then there are $\bar{\mathbf{p}} < \mathbf{c}' = \mathbf{0}'$ which define R by the formula φ_n of Theorem 4.3.*

Proof: We begin with a G which is Cohen 1-generic over C so that $(C \oplus G)' \equiv_T C'$. The set of degrees $\mathcal{R}$ and the finite families of sets of degrees $\mathcal{H}_i$ and $\mathcal{F}_i$ of the proof of Theorem 4.3 are all now uniformly recursive in $C \oplus G$ and consist of pairwise Turing incomparable sets so, by Theorem 4.6, there are sequences of parameters defining each of them all below $(C \oplus G)'$. The proof of Theorem 4.3 now shows that they define R as required. □

We now explain how we plan to code arithmetic in $\mathcal{D}(\leq 0')$. The "intended model" starts with an nice effective successor structure determined by parameters $\bar{\mathbf{q}}$: $\mathbf{c}$, $\mathbf{b}_0$, $\mathbf{b}_1$, $\mathbf{e}_0$, $\mathbf{e}_1$, $\mathbf{d}_0$, $\mathbf{f}_0$ and $\mathbf{f}_1$ with $\mathbf{c}' = 0'$ and $\mathbf{c}$ being above all of the other parameters and all the required $\hat{\mathbf{d}}_n$ as well. Moreover, the $\mathbf{d}_n$ are all uniformly recursive in $\mathbf{c}$. We can do this by Exercise 3.52 or 3.53. We then choose, as in the proof of Theorem 4.4 parameters $\bar{\mathbf{p}}_D$, $\bar{\mathbf{p}}_+$, $\bar{\mathbf{p}}_\times$ and $\bar{\mathbf{p}}_<$ so that $\varphi_1(\bar{\mathbf{p}}_D)$ defines $\{\mathbf{d}_n | n \in \mathbb{N}\}$ and $\varphi_3(\bar{\mathbf{p}}_+)$, $\varphi_3(\bar{\mathbf{p}}_\times)$ and $\varphi_2(\bar{\mathbf{p}}_<)$ (playing the roles of φ_+, $\varphi_\times$ and $\varphi_<$, respectively) that define relations on the countable set defined by $\varphi_1(\bar{\mathbf{p}}_D)$ to determine a structure $\mathcal{M}(\bar{p})$ (where $\bar{\mathbf{p}}$ is the concatenation of all the sequences of parameters used beginning with $\bar{\mathbf{q}}$) that satisfies all the axioms of our finite theory of arithmetic and such that $\mathbf{d}_0$ is the least element in the ordering of $\mathcal{M}(\bar{\mathbf{p}})$ given by $\varphi_2(\bar{\mathbf{p}}_<)$ and, for each n, $\mathbf{d}_{n+1}$ is the immediate successor of $\mathbf{d}_n$ in this order. We can find such parameters below $\mathbf{0}'$ by the arguments for the proof Theorem 4.3 combined with Theorem 4.6 (relativized to $\mathbf{c}$) since the $\mathbf{d}_n$ and the desired relations on them are uniformly recursive in $\mathbf{c}$ and $\mathbf{c}' = \mathbf{0}'$. Now this model is standard since the $\mathbf{d}_n$ are ordered in order type ω and constitute the universe of the model.

The problem is that there is no obvious way to definably say that the universe of the model is precisely the $\mathbf{d}_n$ in terms of just the prescribed parameters (or any other finite list). The issue is that we only have a scheme to generate these degrees not one to define them. We can come fairly close in a first order way. In addition to the correctness conditions that guarantee that the defined relations give a model of arithmetic on $\{x|\varphi_D(x, \bar{\mathbf{p}})\}$, we can approximate niceness by adding the sentences $\mathbf{c} \not\geq \mathbf{b}$ and $\forall d[\varphi_D(d) \rightarrow d \vee \mathbf{c} \geq \mathbf{b} \;\&\; \exists \hat{d}(d \wedge \hat{d} = 0 \;\&\; (\forall d^* \neq d)(\varphi_D(d^*) \rightarrow (d \wedge d^* = \mathbf{0}) \;\&\; (\hat{d} \geq d^*))]$. We can approximate the desired condition that $\{\mathbf{d}_n|n \in \omega\}$ is the domain of our structure by saying that $\mathbf{d}_0$ is the least element in the ordering of $\mathcal{M}(\bar{\mathbf{p}})$ given by $\varphi_2(\bar{\mathbf{p}}_<)$ and for every $\mathbf{d}$ such that $\varphi_D(\mathbf{d}, \bar{\mathbf{p}})$, if $\mathbf{d}$ is an even number in $\mathcal{M}(\bar{\mathbf{p}})$, then $(\mathbf{e}_0 \vee \mathbf{d}) \wedge \mathbf{f}_0$ is its immediate successor in the ordering given by $\varphi_2(\bar{\mathbf{p}}_<)$ while if it is an odd number then its immediate successor is given by $(\mathbf{e}_1 \vee \mathbf{d}) \wedge \mathbf{f}_1$. This guarantees that $\{\mathbf{d}_n|n \in \omega\}$ is the standard part of the model $\mathcal{M}(\bar{\mathbf{p}})$. Thus if we had a formula $\hat{\varphi}_S(x, \bar{r}, \bar{\mathbf{p}})$ which, as $\bar{r}$ ranged over n-tuples from $\mathcal{D}(\leq \mathbf{0}')$, defined a collection of subsets of $\mathcal{M}(\bar{\mathbf{p}})$ that include $\{\mathbf{d}_n|n \in \omega\}$, we could guarantee that $\mathcal{M}(P)$ was standard by saying that every subset (i.e. picked out by some choice of parameters $\bar{r}$) of $\mathcal{M}(\bar{\mathbf{p}})$ which contains its least element $(\mathbf{d}_0)$ and is closed under immediate successor is all of $\mathcal{M}(\bar{\mathbf{p}})$.

The crucial point now is that the proof of Proposition 3.59 shows that, under these conditions, $\{\mathbf{d}_n|n \in \omega\} \in \Sigma_3^C$ as is the ideal generated by this set. That is, the standard part of any $\mathcal{M}(\bar{\mathbf{p}})$ for $\bar{\mathbf{p}}$ satisfying all of these correctness conditions and the ideal it generates are both Σ_3^C. Our goal now is to prove that for every $\mathbf{c} < \mathbf{0}'$ and every Σ_3^C ideal in the degrees below $\mathbf{c}$, there are $\mathbf{g}_0, \mathbf{g}_{,1} \leq_T \mathbf{0}'$ which are an exact pair for the given ideal. Proposition 3.59 and Remark 3.60 then show that we could define the desired set $\{\mathbf{d}_n|n \in \omega\}$ in terms of this exact pair. We latter prove this required result as Theorem 5.12. It supplies the final ingredient of our theorem.

Theorem 4.10: $Th(\mathcal{D} \leq \mathbf{0}') \equiv_{1-1} Th(\mathbb{N})$.

Proof: The above argument (together with Theorem 5.12) shows that we can interpret true first order arithmetic in $\mathcal{D}(\leq \mathbf{0}')$. Thus $Th(\mathbb{N}) \leq_{1-1} Th(\mathcal{D} \leq \mathbf{0}')$. The other direction is immediate since we can define the sets recursive in $0'$ in arithmetic as well as the ordering of Turing reducibility on them. Thus we have a recursive translation of sentences about $\mathcal{D}(\leq \mathbf{0}')$ to ones of arithmetic that preserves truth. Of course, this implies that $Th(\mathcal{D} \leq \mathbf{0}') \leq_{1-1} Th(\mathbb{N})$. □

Notes: Theorem 4.6 and a special case of Theorem 4.9 are in Slaman and Woodin [1986]. The full version of Theorem 4.9 is in Odifreddi and Shore [1991] as is the proof of Theorem 4.10 which is originally due to Shore [1981].

5. Domination Properties

5.1. *Introduction*

An important topic in the study of the complexity of functions from $\mathbb{N}$ to $\mathbb{N}$ is the notion of rate of growth and of one function growing faster than another or faster than a whole class of functions. These issues are not only natural but they have important connections with the computational complexity of the functions as measured by Turing and other reducibilities. In this section we will study some of these notions and their impact on the structure of the degrees. They will play a crucial role in our analysis of the complexity of $D(\leq \mathbf{0}')$. We begin with some basic definitions.

Definition 5.1:

(1) The function g *dominates* the function f ($f < g$) if, for all but finitely many x, $f(x) < g(x)$.
(2) The degree $\mathbf{g}$ *dominates* the function f if some $g \in \mathbf{g}$ dominates f.
(3) The function g *dominates* the degree $\mathbf{f}$ if g dominates every function $f \in \mathbf{f}$.
(4) The degree $\mathbf{g}$ *dominates* the degree $\mathbf{f}$ if for every $f \in \mathbf{f}$ there is a $g \in \mathbf{g}$ which dominates f.

We also sometimes express these relations in the passive form saying, for example, that f is g-*dominated* or f is $\mathbf{g}$-*dominated* for the first two relations. A function g that dominates the degree $\mathbf{0}$ is called *dominant.*

In the literature a degree $\mathbf{f}$ that is not $\mathbf{0}$-dominated (i.e. there is an $f \in \mathbf{f}$ which is not dominated by any recursive function) is, for historical reasons unrelated to our concerns, called *hyperimmune.* If $\mathbf{f}$ is not hyperimmune, i.e. it is $\mathbf{0}$-dominated, is also called *hyperimmune free.* For example, we show later that every $\mathbf{0} < \mathbf{a} < \mathbf{0}'$ is hyperimmune (Theorem 5.4) while the usual minimal degrees constructed below $\mathbf{0}''$ are hyperimmune free.

5.2. *R.E. and* $\mathbf{\Delta}_2^0$ *degrees*

Theorem 5.2: *If $A >_T 0$ is r.e. then there is a function $m \equiv_T A$ which is not* $\mathbf{0}$*-dominated, i.e. it is not dominated by any recursive function. Indeed, any function g which dominates m computes A.*

Proof: For A r.e., let A_s be the standard approximation to A at stage s. Let m be the least modulus function for this approximation: $m(x) = \mu s(\forall t \geq s)(A_s \upharpoonright x = A_t \upharpoonright x)$. For r.e. sets, the approximation changes its mind at most once and is correct in the limit, so $m(x)$ is also the $\mu s(A_s \upharpoonright x = A \upharpoonright x)$ and is clearly of the same degree as A. Moreover, if $g(x) \geq m(x)$ for almost all x, then $A \leq_T g$ as $A \upharpoonright x = A_{g(x)} \upharpoonright x$ for all but finitely many x. Thus, if $A >_T 0$, then m is not dominated by any recursive function and any g that dominates m computes A. □

The Shoenfield limit lemma (Lemma 1.11) gives us a recursive approximation $h(x, s)$ to any $A \in \Delta_2^0$ (or equivalently $A \leq_T 0'$). So the least modulus function m makes sense for such an approximation as well. So does the second version used in the above proof. Here we call it the *computation function:* $f(x) = \mu(s > x)(\forall y < x)(h(y, s) = A(y))$ (for technical reasons, we do not consider first few stages). It calculates the first stage after x at which the approximation is correct up to x. But, since we are no longer looking at r.e. sets, the approximation might change even after it's correct and the computation function f need not be the same as the least modulus m. The two functions may not be the same even up to degree.

Exercise 5.3: Find an $A <_T 0'$ and an approximation $h(x, s)$ to A for which the least modulus function m computes $0'$. On the other hand, the computation function f for h is always of the same degree as A.

We can, nonetheless extend Theorem 5.2 to all $A \in \Delta_2^0$.

Theorem 5.4: *If A is Δ_2^0, then there is an $f \equiv_T A$ which is not* $\mathbf{0}$*-dominated. Indeed, any function g which dominates f computes* $\mathbf{a}$.

Proof: By the Shoenfield limit lemma, there is a recursive $h(x, s)$ such that $\lim_{s \to \infty} h(x, s) = A(x)$. Let $f(x)$ be the computation function for this approximation. Suppose $f < g$. We claim that even though $h(z, s)$ may change at $z < x$ for $s > f(x)$, we can still compute A from g. Let s_0 be such that $(\forall m \geq s_0)(f(m) < g(m))$. To calculate $A(n)$ for $n > s_0$ find an $s > n$ such that $h(n,t)$ is constant for $t \in [g(s), gg(s)]$. Since $h(n,t)$ is eventually constant, such an s exists. Moreover, we can find it recursively in g: compute the intervals $[g(n+1), gg(n+1)], [g(n+2), gg(n+2)], [g(n+3), gg(n+3)], \ldots$ checking to see if h is constant on the intervals. By the clause that makes $f(x) > x$ in the definition of the computation function and our choice of s_0, $gg(s) > fg(s) > g(s)$, so the first $t > g(s)$ at which h is correct for all elements below $g(s)$ is in $[g(s), gg(s)]$. For this t, $h(n,t) = A(n)$. As we chose s so that the value of $h(n,t)$ is constant on this interval, $A(n) = h(n,t)$ for any $t \in [g(s), gg(s)]$ so we have computed A recursively in g as required. □

Exercise 5.5: What are the correct relativizations of the previous two theorems?

Exercise 5.6: The above results can be extended by iterating the notions of "r.e. in" or more generally "Δ^0_2 in" as long as one includes the lower degrees. We say that A 1-REA if it is r.e. then we define n-REA by induction: A is $n+1$-REA if A is of the form $B \oplus W^B_e$ where B is n-REA. (REA stands for r.e. in and above.) Prove that any n-REA set A has an $f \equiv_T A$ such that any $g > f$ computes A. Do the same with Δ^0_2 replacing r.e. These results can be carried into the transfinite. Prove, for example, that $0^{(\omega)}$ has the same property.

Theorem 5.7: *If $A > 0$ is r.e. and $\mathcal{P}$ is a recursive notion of forcing then there is a 1-generic sequence $\langle p_s \rangle \leq_T A$ so that the corresponding 1-generic G is recursive in A as well.*

Proof: We build a 1-generic sequence p_s recursive in A. Let $f <_T A$ be the least modulus function for A. Let S_e be the eth Σ_1 set of conditions. The requirements are

$$R_e : \text{for some } s,\ p_s \in S_e \text{ or } (\forall q \leq p_s)(q \notin S_e).$$

At stage s, we have a condition p_s. Note that we are thinking of P as a subset of $\mathbb{N}$ and so have the natural ordering $\leq$ on its members (and all of $\mathbb{N}$) as well as the forcing ordering $\leq_{\mathcal{P}}$. We say that R_e has been declared satisfied by stage s if there is a p_n with $n \leq s$ such that $p_n \in S_{e,f(s)}$. Find the least $e < s$ such that R_e has not yet been declared satisfied and such that $(\exists q \leq_{\mathcal{P}} p_s)(q \leq f(s)\ \&\ q \in S_{e,f(s)})$. For this e, choose the least such q and put $p_{s+1} = q$. If there is no such e, let $p_{s+1} = p_s$.

To verify that the construction succeeds, suppose for the sake of a contradiction that e_0 is least such that

$$\neg\exists s(p_s \in S_{e_0} \vee (\forall q \leq_{\mathcal{P}} p_s)(q \notin S_{e_0})).$$

Choose $s_0 > e_0$ such that $\forall i < e_0$ if there is a $p_s \in S_i$ then there is one with $s < s_0$ and $p_s \in S_{i,f(s_0)}$ (so by this stage we have already declared satisfied all higher priority requirements that are ever so declared). We claim that we can now recursively recover the entire construction and the values of $f(s)$ for $s \geq s_0$. As this would compute A recursively, we would have our desired contradiction. Consider what happens in the construction at each stage $s \geq s_0$ in turn. Suppose we have p_s. At stage s we look for the least $e < s$ such that $(\exists q \leq_{\mathcal{P}} p_s)(q \leq f(s) \,\&\, q \in S_{e,f(s)})$. There is no such $e < e_0$ by our choice of s_0. If e_0 itself were such an e, we would act for it and declare P_{e_0} to be satisfied, contrary to our choice of e_0. On the other hand, by our choice of e_0 there is a $q \leq_{\mathcal{P}} p_s$ with $q \in S_{e_0}$. We can find such a q recursively (because we know it exists). We did not find this q in the construction at stage s because either $q > f(s)$ or $q \in S_{e_0} - S_{e_0,f(s)}$. So we can now find a bound t on $f(s)$ by finding the stage at which q enters S_{e_0}. Given $t \geq f(s)$ we can calculate $f(s)$ as the least z such that $A_z \upharpoonright s = A_t \upharpoonright s$. Once we have $f(s)$ we can recursively determine what happened at stage s of the construction and in particular the value of p_{s+1}. Thus we can continue our recursive computation of $f(s)$ as claimed. □

Relativizing Theorem 5.7 to C gives, for any C recursive notion of forcing $\mathcal{P}$, a $G \leq_T A$ which is C 1-generic for $\mathcal{P}$ for any $A >_T C$ which is r.e. in C.

Exercise 5.8: The crucial property of the function f used in the above construction was that there is a uniformly recursive function computing $f(x)$ from any number greater than it. Prove that if there is a partial recursive $\varphi(x, s)$ such that $(\forall s \geq f(x))(\varphi(x, s) = f(x))$ then f is of r.e. degree.

Corollary 5.9: *If* $\mathbf{a} > \mathbf{0}$ *is r.e. then there is Cohen 1-generic* $G <_T A$ *and so, for example, every countable partial order can be embedded in the degrees below* $\mathbf{a}$.

Similarly we have

Corollary 5.10: *If* $\mathbf{a}$ *is r.e. in* $\mathbf{b}$ *and strictly above it, then every partial lattice recursive in* $\mathbf{b}$ *can be embedded into* $[\mathbf{b}, \mathbf{a})$.

Corollary 5.11: *If* $\mathbf{a}$ *is r.e. then every maximal chain in* $(\mathcal{D}(\leq \mathbf{a}), \leq_T)$ *is infinite. In fact, there is no maximal element less than* $\mathbf{a}$ *in* $(\mathcal{D}(\leq \mathbf{a}), \leq_T)$.

Proof: Suppose $\mathbf{b} < \mathbf{a}$. Then $\mathbf{a}$ is r.e. in and strictly above $\mathbf{b}$. Relativizing Theorem 5.7 to a $B \in \mathbf{b}$ and using Cohen forcing gives us a $G \leq_T A$ which is Cohen 1-generic over B. So the degrees of $B \oplus G^{[i]}$ are in fact all between $\mathbf{b}$ and $\mathbf{a}$ and even independent. □

We now apply Theorem 5.7 to provide the missing way of identifying the standard parts of effective successor models coded below $0'$ that we need to calculate the complexity of $Th(\mathcal{D}(\leq \mathbf{0}'))$.

Theorem 5.12: *If* $A >_T C$, A *is r.e. in* C *and* I *is an ideal in* $\mathcal{D}(\leq \deg(C))$ *such that* $W = \{e : \deg(\Phi_e^C) \in I\} \in \Sigma_3^C$ *then there is an exact pair* G_0, G_1 *for* I *below* A.

Proof: We provide a C-recursive notion of forcing $\mathcal{P}$ such that any 1-generic for $\mathcal{P}$ gives an exact pair for I and apply Theorem 5.7 relativized to C. The conditions of $\mathcal{P}$ are of the form $p = \langle p_0, p_1, F_p, n_p \rangle$ where $p_i \in 2^{<\omega}$, $|p_0| = |p_1| = |p|$, $F_p \in \omega^{<\omega}$, $n_p \in \omega$ such that

$$(\forall i \in \{0,1\})(\forall \langle e,x,y \rangle)(\exists^{\leq 1} \langle w,m \rangle)\left(\langle e,x,y,w,m \rangle \in p_i\right).$$

We define V as expected $V(p) = p_0 \oplus p_1$. So for a 1-generic $\mathcal{G}$, we have $G_i = \cup\{p_i | p \in \mathcal{G}\}$. If $e \in W$, we want Φ_e^C to be coded into G_i. The unusual restriction above on conditions in P suggests how we intend to do this coding. Since $W \in \Sigma_3^C$ we have a relation $R \leq_T C$ such that $e \in W \Leftrightarrow \exists x \forall y \exists z R(e,x,y,z)$. We denote the pairs of elements of W and their witnesses by $\hat{W} = \{\langle e,x \rangle : \forall y \exists z R(e,x,y,z)$. To calculate Φ_e^C for $e \in W$, our plan is to first choose an x such that $\langle e,x \rangle \in \hat{W}$. We then search for $\langle w,m \rangle$ such that $\langle e,x,y,w,m \rangle \in G_i$ and announce that $\Phi_e^C(y) = m$. The definition of P guarantees that this procedure gives at most one answer. The definition of the partial order $\leq_{\mathcal{P}}$ below guarantees that this procedure makes only finitely many mistakes for any 1-generic. Genericity also guarantees that, when $\langle e,x \rangle \in \hat{W}$, it gives a total function.

The number n_p in our conditions acts as a bound for how far we have to search to sufficiently verify the Π_2 assertion that x is a witness that $e \in W$ (and so also that Φ_e^C is total). The set F_p tells us for which $\langle e,x \rangle$ we can make no further mistakes in our coding of Φ_e^C into $G_i^{\langle e,x \rangle}$ when we extend p. With this intuition, we define extension in $\mathcal{P}$ by $q \leq_{\mathcal{P}} p$ iff

$$q_i \supseteq p_i, \qquad F_q \supseteq F_p, \qquad n_q \geq n_p,$$

and

$$(\forall i \in \{0,1\})(\forall \langle e,x,y,w,m\rangle \in [|p|,|q|)(\langle e,x\rangle \in F_p \ \& \ \langle e,x,y,w,m\rangle \in q_i \\ \rightarrow \Phi^C_{e,n_q}(y) = m \ \& \ \forall y' \leq y \exists z \leq n_q\,(R(e,x,y',z))\,.$$

Note that $\mathcal{P}$ is recursive in C.

Suppose that G_0, G_1 are given by a C-1-generic sequence $\langle p_s\rangle \leq_T A$ as in Theorem 5.7 relativized to C. We claim that G_0, G_1 are an exact pair for I.

First assume that $\langle e,x\rangle \in \hat{W}$. We show that $\Phi^C_e \leq_T G_i$. As the sets $\{p|\,\langle e,x\rangle \in F_p\}$ are obviously dense in $\mathcal{P}$, there is an s such that $\langle e,x\rangle \in F_{p_s}$. For any $\langle e,x,y,w,m\rangle \in p_t$ with $t > s$, $\Phi^C_e(y) = m$ by definition and so as noted above, the prescribed search procedure which is recursive in G_i returns only correct answers for $y > |p_s|$. Next, we claim that for each $y > |p_s|$, $i \in \{0,1\}$ and $m = \Phi^C_e(y)$ the Σ^C_1 sets $S_{e,x,y,m,i} = \{r|\exists w(\langle e,x,y,w,m\rangle \in r_i\}$ are dense below p_s. This guarantees that $\langle p_t\rangle$ meets each of these sets and so the search procedures are total and correctly compute $\Phi^C_e(x)$ for all but finitely many x. To see that these sets are dense below p_s, consider any $q \leq p_s$ with no w such that $\langle e,x,y,w,m\rangle \in q_i$. Choose any $w > |q|$ and define an $r \leq_{\mathcal{P}} q$ by making $|r| = \big\langle e,x,y,w,\Phi^C_e(y)\rangle + 1\big\rangle$, $r_i = q_i \cup \{\langle e,x,y,w,\Phi^C_e(y)\rangle\}$ (i.e. we let them be 0 at other points below the length), $F_r = F_q$ and letting n_r be the least $n \geq n_q$ such that $\forall y' \leq y \exists z < n(R(e,x,y',z) \ \& \ \Phi^C_{e,n}(y)\downarrow)$ (one such exists since we are assuming that $\langle e,x\rangle \in \hat{W}$). Then $r \leq_{\mathcal{P}} q$ and $r \in S_{e,x,y,m,i}$ as desired.

We next want to deal with the minimality conditions associated with the G_i being an exact pair for I. Suppose then that $\Phi^{G_0}_e = \Phi^{G_1}_e = D$ is total. We want to prove that $D \leq \oplus\{\Phi^C_e : e \in F\}$ for some finite $F \subset W$. Consider the Σ_1 set S_e of conditions p:

$$S_e = \{p : \exists n\,(\Phi^{p_0}_e(n)\downarrow \neq \Phi^{p_1}_e(n))\downarrow\}.$$

By our assumption there is no $p_s \in S_e$ so we have a $p_s = p$ such that $\forall q \leq_{\mathcal{P}} p(q \notin S_e)$. We claim that $D \leq \oplus\{\Phi^C_e : \langle e,x\rangle \in F_p \cap \hat{W}\}$. For every $\langle e,x\rangle \in F_p \backslash \hat{W}$, let $y(e,x)$ be the least y such that $\neg\forall y' \leq y \exists z R(e,x,y',z) \vee \Phi^C_e(y)\uparrow$. It is clear that there is no $q \leq_{\mathcal{P}} p$ with any $\langle e,x,y,w,m\rangle \in q_i$ for $\langle e,x\rangle \in F_p \setminus \hat{W}$ and $y \geq y(e,x)$. Choose $q \leq_{\mathcal{P}} p$ in $\langle p_s\rangle$ so that it has the maximal number of y's with some $\langle e,x,y,w,m\rangle \in q_i$ for $y < y(e,x)$ and $i \in \{0,1\}$. To compute $D(y)$ for $y > |q|$, we find a $t \in \mathcal{P}$ such that $t_i \supseteq q_i$, $\Phi^{t_0}_e(y)\downarrow = \Phi^{t_1}_e(y)\downarrow$, no elements not in q_i are added into t_i in columns $\langle e,x\rangle \in F_p \backslash \hat{W}$ and for any $\langle e,x,y,w,m\rangle \in t_i$ with $\langle e,x\rangle \in F_p \cap \hat{W}$,

$\Phi_e^C(y) = m$. Such an extension exists because $\Phi_e^{G_0}(y) \downarrow = \Phi_e^{G_1}(y) \downarrow$ and by the maximality property of q and the definition of $\leq_{\mathcal{P}}$, $G_i^{[\langle e,x\rangle]} = q_i^{[\langle e,x\rangle]}$ for $\langle e,x\rangle \in F_p \setminus \hat{W}$ and so there is such a $\hat{t} \in \langle p_s\rangle$. Finding one such t is clearly recursive in $\oplus\{\Phi_e^C : \langle e,x\rangle \in F_p \cap \hat{W}\}$. Thus we only need to show that any such t provides the right answer. If one such gave an answer different than that given by $\hat{t}$ (and so G_0 and G_1) then $\langle t_0, \hat{t}_1, F_p, n\rangle$ (where $n \geq n_q$ is large enough so that $\Phi_{e,n}^C(y) \downarrow$ for every $\langle e,x,y,w,m\rangle$ in t_0 or $\hat{t}_1$ with $\langle e,x\rangle \in F_p \cap \hat{W}$) would be an extension of p in S_e for the desired contradiction. □

This Theorem completes the proof of Theorem 4.10 that the theory of the degrees below $\mathbf{0}'$ is recursively isomorphic to true arithmetic. We can extend the result to all r.e. degrees.

Exercise 5.13: For every r.e. $\mathbf{r} > \mathbf{0}$, $Th(\mathcal{D}(\leq\mathbf{r}) \equiv_{1-1} Th(\mathbb{N})$.

Notes: Theorem 5.2 is due to Dekker [1954]; Theorem 5.4 to Miller and Martin [1968]. We are not sure who first proved Corollary 5.9 (presumably using a different method called r.e. permitting). The style of proof based directly on domination properties used here to prove Theorem 5.7 is attributed to us in Soare [1987, Ch. VI Exercise 3.9] in the case of Cohen forcing. Theorem 5.12 is in Shore [1981] which also is the original source of Exercise 5.13.

5.3. *High and $\overline{GL_2}$ degrees*

We now look at stronger domination properties and their relation to the jump classes $\mathbf{H}_1$ and $\mathbf{L}_2$ below $\mathbf{0}'$ and their generalizations. Recall from Definition 1.12 that for $\mathbf{a} \leq \mathbf{0}'$, $\mathbf{a} \in \mathbf{H}_1 \Leftrightarrow \mathbf{a}' = \mathbf{0}''$; $\mathbf{a} \in \mathbf{L}_2 \Leftrightarrow \mathbf{a}'' = \mathbf{0}''$. For degrees $\mathbf{a}$ not necessarily below $\mathbf{0}'$, $\mathbf{a} \in \mathbf{GL}_2 \Leftrightarrow (\mathbf{a} \vee \mathbf{0}')' = \mathbf{a}''$; $\mathbf{a} \in \mathbf{GH}_1 \Leftrightarrow \mathbf{a}' = (\mathbf{a} \vee \mathbf{0}')'$. It is also common to say that $\mathbf{a}$ is *high* if $\mathbf{a}' \geq \mathbf{0}''$. As it turns out these last are the degrees of dominant functions. Of course, $\mathbf{a} \in \overline{\mathbf{GL}_2}$ means that $\mathbf{a} \notin \mathbf{GL}_2$. We relativize these notions to degrees above $\mathbf{b}$ by writing, for example, $\mathbf{a} \in \overline{\mathbf{GL}_2}(\mathbf{b})$.

Let us begin by showing that there is there a dominant function. In fact, if $\mathcal{C}$ is any countable class of functions $\{f_i\}$ then there is function f which dominates all the f_i. For example, put $f(x) = \max\{f_i(x) : i < x\} + 1$. This construction requires a uniform list of all the functions f_i. For the recursive functions we know that $0''$ can compute such a list: $Tot = \{e : \Phi_e \text{ total}\}$

is clearly Π_2^0 and so recursive in $0''$ by the Hierarchy Theorem (Theorem 1.10) and so there is a sequence f_i uniformly computable from $0''$ which then computes a dominant function as described. We can do better than this and avoid using totality. If $f(x) = \max\{\Phi_e(x) : e < x \ \& \ \Phi_e(x)\downarrow\}$ then $f \leq_T 0'$ and is also clearly dominant. We can even do a bit better and get away with functions of high degree.

Theorem 5.14: ***(Martin's High Domination Theorem)*** *A set A computes a dominant function f if and only if $0'' \leq_T A'$.*

Proof: Suppose first that $0'' \leq_T A'$. By the Shoenfield limit lemma (Theorem 1.11) and the fact that $Tot \leq_T 0''$, there is an $h \leq_T A$ with $\lim_{s\to\infty} h(e,s) = Tot(e)$. We want to compute a function f recursively in A such that, for every e for which Φ_e is total, $f(x)$ is larger than $\Phi_e(x)$ for all but finitely many x. Any such f is dominant. To compute $f(x)$ we compute, for each $e < x$, both $\Phi_{e,t}(x)$ and $h(e,t)$ for $t \geq x$ until either the first one converges, say to y_e, or $h(e,t) = 0$. As, if Φ_e is not total, $\lim h(e,t) = 0$, one of these outcomes must happen. We set $f(x)$ to be one more than the maximum of all the y_e so computed for $e < x$. Note that $f \leq_T h \leq_T A$. It remains to verify that if Φ_e is total then $\Phi_e < f$. By our choice of h, $\exists s_0(\forall s \geq s_0)(h(e,s) = 1)$. So for $x > s_0$ when we calculate $f(x)$ we always find a t such that $\Phi_{e,t}(x)\downarrow = y_e$ and so $f(x) > \Phi_e(x)$ for all $x > s_0$. For the other direction, suppose we have a dominant f. As Tot is Π_2^0 and computes $0''$, it suffices to show that it is also $\Sigma_2(f)$ as it would then be $\Delta_2(f)$ and so recursive in f'. We claim that

$$\forall x \exists s \Phi_{e,s}(x)\downarrow \;\Leftrightarrow\; \exists c \forall x \Phi_{e,f(x)+c}(x)\downarrow .$$

Suppose Φ_e is total (if not, then of course both conditions fail). Let $k(x) = \mu s \Phi_{k,s}(x)\downarrow$. Then k is recursive (because we know that $\forall x \Phi_e(x)\downarrow$). By hypothesis, f dominates k. Thus, the right hand side holds. This is a $\Sigma_2(f)$ formula as desired. □

Now a look at the definitions shows that for $\mathbf{a} \leq_T \mathbf{0}'$, $\mathbf{a} \notin \mathbf{L}_2$ is equivalent to $\mathbf{0}'$ not being high relative to $\mathbf{a}$. Relativizing Theorem 5.14 to an $\mathbf{a} \leq_T \mathbf{0}'$ we see that $\mathbf{a} \notin \mathbf{L}_2$ if and only if no $f \leq_T 0'$ dominates every (total) function recursive in A. We can then handle $\overline{\mathbf{GL}}_2$ by relativizing to $\mathbf{a} \vee \mathbf{0}'$ to prove the following:

Proposition 5.15: *A set $A \leq_T 0'$ has degree in $\overline{\mathbf{L}}_2$ if and only if $(\forall g \leq_T 0')(\exists f \leq_T A)(f \not< g)$. An arbitrary set A has degree in $\overline{\mathbf{GL}}_2$ if and only if $(\forall g \leq_T A \vee 0')(\exists f \leq_T A)(f \not< g)$.*

Proof: Exercise. □

This says that, while sets that are not high do not compute dominant functions, if they are not too low they compute functions which are not dominated by any recursive function. This suffices for many applications.

Theorem 5.16: *If $A \notin GL_2$ then for any recursive notion of forcing $\mathcal{P}$ there is 1-generic sequence $\langle p_s \rangle \leq_T A$ and so the associated 1-generic G is also recursive in A.*

Proof: For any $g \leq_T A \vee 0'$, there is an $f \leq_T A$ not dominated by g. Without loss of generality we may take f to be strictly increasing. We first construct the function g that we want and then, using the associated f, we construct a 1-generic sequence p_s recursively in f (and so A). We again make use of the natural order $\leq$ on $P \subseteq \mathbb{N}$.

Let S_e list the Σ_1 subsets of P. As usual, we declare S_e to be satisfied at s if $(\exists n \leq s)(p_n \in S_{e,s})$. We define g by recursion using $0'$. Given $g(s)$, we want to determine $g(s+1)$. For each condition $p \leq g(s)+1$, ask $0'$ if $(\exists q \leq_{\mathcal{P}} p)(q \in S_e)$ for each $e \leq g(s)+1$. If such an extension exists, let x_e be the least x such that $(\exists q \leq_{\mathcal{P}} p)(q \leq x \;\&\; q \in S_{e,x})$. Put $g(s+1) = \max\{x_e | e \leq g(s)+1\}$.

We cannot use g itself in the construction of the desired 1-generic $\langle p_s \rangle$ because we want $\langle p_s \rangle \leq_T A$. But, since $g \leq_T A \vee 0'$, we can use an increasing $f \leq_T A$ not dominated by g. The construction of G is recursive in f (hence in A). At stage s, we have finite a condition p_s. For each $e \leq s$ not declared satisfied at s, see if $(\exists q \leq_{\mathcal{P}} p_s)(q < f(s+1) \;\&\; q \in S_{e,f(s+1)})$. If so, take the smallest such q for the least such e and let it be p_{s+1}. If not, $p_{s+1} = p_s$. The construction is recursive in f, hence in A. Thus $\langle p_s \rangle \leq_T A$ and the associated $G \leq_T A$ as well. Note that $p_s \leq f(s)$ by induction. Indeed $p_s \leq g(s)$ as well because $g(s)$ gives a bound on the witness required in the definition of p_s.

To verify that G is 1-generic suppose, for the sake of a contradiction, that there is a least e_0 such that

$$\neg \exists s (p_s \in S_{e_0} \vee (\forall p \leq_{\mathcal{P}} p_s)(p \notin S_{e_0})).$$

Choose s_0 such that, $(\forall i < e_0)[(\exists s)(S_i$ is declared satisfied at $s) \rightarrow S_i$ is declared satisfied by $s_0]$. Consider any $s > s_0$ at which $f(s+1) > g(s+1)$. By our choice of e_0, there is a $q \leq_{\mathcal{P}} p_s$ such that $q \in S_{e_0}$. Moreover, as $p_s \leq g(s)$, by definition of g there is one $\leq g(s+1)$ such that it belongs to $S_{e_0, g(s+1)}$ as well. By our choice of s, $q \leq g(s+1) < f(s+1)$. Thus at stage $s+1$, we would act to extend p_s to a $p_{s+1} \in S_{e_0}$ for the desired contradiction. □

As for the r.e. degrees, having a 1-generic below a degree $\mathbf{a} \notin \mathbf{GL}_2$ provides a lot of information about the degrees below $\mathbf{a}$. For example, as in Corollary 5.9, we can embed every countable partial order below any $\mathbf{a} \notin \mathbf{GL}_2$. It is tempting to think that we could also prove the analog of Corollary 5.11 that every maximal chain in the degrees below $\mathbf{a}$ is infinite. This is true for $\mathbf{a} < \mathbf{0}'$ (Exercise 5.17) but was a long open question (Lerman [1983]). Cai [2012] has now proven that it is not true. There are $\mathbf{a} \notin \mathbf{GL}_2$ which are the tops of a maximal chain of length three.

Exercise 5.17: Prove that if $\mathbf{a} \leq \mathbf{0}'$ and $\mathbf{a} \notin \mathbf{L}_2$ then any maximal chain in the degrees below $\mathbf{a}$ is infinite.

On the other hand, we can say quite a bit that is not true of arbitrary r.e. degrees about the degrees above $\mathbf{a}$ when $\mathbf{a} \notin \mathbf{GL}_2$.

Definition 5.18: A degree $\mathbf{a}$ has the *cupping property* if $(\forall \mathbf{c} > \mathbf{a})(\exists b < c)$ $(\mathbf{a} \vee \mathbf{b} = \mathbf{c})$.

Theorem 5.19: *If* $\mathbf{a} \in \overline{\mathbf{GL}_2}$ *then* $\mathbf{a}$ *has the cupping property. Indeed, if* $A \notin GL_2$ *and* $C >_T A$ *then there is* $G \not\geq_T A$ *such that* $A \vee G \equiv_T C$ *and* G *is Cohen* 1*-generic.*

Proof: We need to add requirements $R_e : \Phi_e^G \neq A$ to the proof of Theorem 5.16 for Cohen forcing (making all the requirements into a single list Q_e) and code C into G as well (so as to be recoverable from $A \oplus G$). In the definition of $g(s+1)$ in that proof, for each $p \leq g(s)+1$ look as well for $q_0, q_1 \supseteq p$ and x such that $q_0|_e q_1$. Then make $g(s+1)$ also bound the least such extensions τ_0, τ_1 for each $e, p \leq g(s)+1$ for which such extensions exist.

Again choose $f \leq_T A$ strictly increasing and not dominated by g. The construction is done recursively in $f \oplus C$. At stage s we have p_s and we look for the least e such that Q_e has not yet been declared satisfied and for which there is either a $q \leq_{\mathcal{P}} p$ with $q \leq f(s+1)$ that would satisfy Q_e as before if it is an S_i or a pair of strings $q_0, q_1 \supseteq p_s$ with $q_i \leq f(s+1)$ such that $q_0|_e q_1$ if $Q_e = R_i$. Let e be the least for which there are such extensions. If $Q_e = S_i$ choose q as before. If it is R_i Let q be the q_j such that $\Phi_e^{q_j}(x) \downarrow \neq A(x)$. We then let $p_{s+1} = q\hat{\ }C(s)$ and declare Q_e to be satisfied. If there is no such e, we let $p_{s+1} = p_s\hat{\ }C(s)$. Note that $p_{s+1} \leq f(s+1)+1$ (the extra 1 comes from appending $C(s)$).

Since the construction is recursive in $f \oplus C$ and $f \leq_T A \leq_T C$, we have $G \leq_T C$. But, $C \leq_T \langle p_s \rangle$ because $C(s) = p_{s+1}(|p_{s+1}|)$. However, $\langle p_s \rangle \leq_T A \oplus G$ because $f \leq_T A$ tells how to compute each stage from the given p_s to the choice of q. Then G tells us the last extra bit at the end of p_{s+1}.

To verify that G has the other required properties suppose e_0 is least such that Q_e fails. Assume that by stage s_0 we have declared all requirements with $e' < e_0$ which will ever be declared satisfied to be satisfied. Consider a stage $s > s_0$ at which $f(s+1) > g(s+1)$. If $Q_e = S_i$ then we argue as in the previous theorem. If $Q_e = R_i$ and there were any $q_0, q_1 \supseteq p_s$ with $q_0|_e q_1$ then would have taken one of them as our q and declared $Q_e = R_i$ to be satisfied contrary to our choice of e_0. On the other hand, if there are no such extensions, then as usual Φ_e^G is recursive if total and so R_i would also succeed contrary to our assumption. □

Remark 5.20: Not every r.e. degree has the cupping property.

For other results about $\overline{\mathbf{GL}_2}$ degrees it is often useful to strengthen Theorem 5.16 to deal with notions of forcing recursive in A rather than just recursive ones.

Theorem 5.21: *For $A \in \overline{GL_2}$, given an A recursive notion of forcing $\mathcal{P}$ and a sequence D_n of dense sets uniformly recursive in $A \vee 0'$ (or with a density function $d(n,p) \leq_T A \vee 0'$) there is a generic sequence $\langle p_s \rangle \leq_T A$ meeting all the D_n. Of course, the generic G associated with the sequence is recursive in A as well.*

Proof: Let m_K be the least modulus function for $K = 0'$ and let $\Psi_n^{A \oplus K} = D_n$, i.e. the Ψ_n uniformly compute membership in D_n. We define $g \leq_T A \vee 0'$ by recursion. Given $g(s)$ we find, for each $p, n \leq g(s)+1$ the least q such that $q \leq_P p$ and $q \in D_n$ as witnessed by a computations of $\Psi_{n,u}^{A \oplus K_u \restriction u}(n) = 1$ where K_u is the same as K on the use from K in this computation. Next we let $g(s+1)$ be the least number larger than q, u and $m_K(u)$ for all of these q and u as well as $m_K(g(s)+1)$. As $g \leq_T A \vee 0'$ and $A \in \overline{GL_2}$ there is an increasing $f \leq_T A$ not dominated by g.

We construct the sequence $\langle p_s \rangle$ recursively in $f \leq_T A$. At stage s we have p_s. Our plan is to satisfy the requirement of meeting D_n for the least n for which we do not seem to have done so yet and for which we can find an appropriate extension of p_s when we restrict our search to $q \leq f(s+1)$ as well as our use of $0'$ to what we have at stage $f(s+1)$. More formally, we determine (recursively in A) for which D_n $(n \leq s)$ there is a $t \leq s$ such that $\Psi_n^{(A \oplus K_{f(s+1)}) \upharpoonright f(s+1)}(p_t) = 1$. Among the other $n \leq s$, we search (again recursively in A) for one such that $(\exists q \leq_{\mathcal{P}} p_s)(q \leq f(s+1)$ & $\Psi_n^{(A \oplus K_{f(s+1)}) \upharpoonright f(s+1)}(q) = 1)$. If there is one we act for the least such n by letting p_{s+1} be the least such q for this n. If not, let $p_{s+1} = p_s$. Note that $p_{s+1} \leq f(s+1)$ by the restriction on the search space and $p_{s+1} \leq g(s+1)$ as well since $g(s+1)$ also bounds the least witness by the definition of g.

We now claim that for each n there is a $p_s \in D_n$. If not, suppose, for the sake of a contradiction, that n is the least counterexample. Choose s_0 such that for all $m < n$ there is $t < s_0$ such that $p_t \in D_m$ and indeed such that $\Psi_m^{(A \oplus K_{s_0}) \upharpoonright s_0}(p_t) = 1$ and $K_{s_0} \upharpoonright u = K \upharpoonright u$ where u is the use of this computation of Ψ_m at p_t. Thus, by construction, we never act for $m < n$ after s_0. As g does not dominate f we may choose an $s > s_0$ with $f(s+1) > g(s+1)$. At stage s we have p_s and $p_t \notin D_n$ for all $t \leq s$ in the sense required, i.e. $\Psi_n^{(A \oplus K_{f(s+1)}) \upharpoonright f(s+1)}(p_t) = 0$ since any computation of this form gives the correct answer by our definition of $g(s+1)$ and the fact that $f(s+1) > g(s+1)$. There is a $q \leq_{\mathcal{P}} p_s$ with $q \in D_n$ and the least such is less than $f(s+1)$ and $\Psi_n^{(A \oplus K_{f(s+1)}) \upharpoonright f(s+1)}(q) = 1$ with the computation being a correct one from $A \oplus K$ by the definition of $g(s+1) < f(s+1)$. Thus we would take the least such q to be $p_{s+1} \in D_n$ for the desired contradiction. □

We now give a couple of applications that play a crucial role in our global analysis of definability in $\mathcal{D}(\leq \mathbf{0}')$. The first is a jump inversion theorem that generalizes Shoenfield's (Corollary 5.23).

Theorem 5.22: *($\overline{\mathbf{GL}}_2$ **Jump Inversion**) If $A \in \overline{GL}_2$, $C \geq_T A \vee 0'$, and C is r.e. in A, then there is a $B \leq_T A$ such that $B' \equiv_T C$.*

Proof: Let C_s be an enumeration of C recursive in A. We want a notion forcing recursive in A and a collection of dense sets D_n such that for any $\langle D_n \rangle$ generic G, $G' \equiv_T C$. This time, our notion of forcing has conditions $p \in 2^{<\omega}$. The definition of extension for $\mathcal{P}$ is a bit tricky. If $q \supseteq p$ and

$$\langle e, x \rangle \in [|p|, |q|) \Rightarrow [C_{|p|}(x) = q(\langle e, x \rangle) \text{ or } \exists n \leq e\, (\Phi_n^p(n)\uparrow \ \&\ \Phi_n^q(n)\downarrow)]$$

we say that $q \leq_1 p$. Now this relation is clearly recursive in A since A computes $C_{|p|}$ for each p. However, it need not be transitive (Exercise). We let $\leq_{\mathcal{P}}$ be its transitive closure. As, given any $r \supseteq p$, there are only finitely many q's with $r \supseteq q \supseteq p$ we can check all possible routes via $\leq_1$ from p to r recursively in A and so $\leq_{\mathcal{P}}$ is also recursive in A. The plan for coding C into G' uses the Shoenfield limit lemma 1.11 and partially explains the notion of extension. It guarantees that $e \in C \Rightarrow G^{[e]} =^* \omega$ while $e \notin C \Rightarrow G^{[e]} =^* \emptyset$. Thus $e \in C \Leftrightarrow \lim_s G(\langle e, s \rangle = 1$ and so $C \leq_T G'$. Suppose we have a generic sequence $\langle p_s \rangle \leq_T A$ for some collection of dense sets as in Theorem 5.21. The definition of extension guarantees that coding mistakes can happen in column e only when $\Phi_n^{p_s}(n)$ first converges for some $n \leq e$. Thus $C \leq_T G'$.

Our first class of dense sets include the trivial requirements and in addition force the jump of G in the hope of making $G' \leq_T C$:

$$D_{m,j} = \{p : |p| \geq j \ \& \ [\Phi_m^p(m) \downarrow \text{ or } (\forall q \supseteq p)(\Phi_m^q(m) \uparrow \text{ or } (\exists e < m)$$
$$(\exists \langle e, x \rangle \in [|p|, |q|)(C_{|p|}(e) \neq q(\langle e, x \rangle) \text{ but } \neg(\exists n \leq e)(\Phi_n^p(n) \uparrow \ \& \ \Phi_n^q(n) \downarrow))]\}$$

Note that, after we use A to compute $C_{|p|}$, membership in $D_{m,j}$ is a Π_1 property and so recursive in $0'$. Thus, the $D_{m,j}$ are uniformly recursive in $A \vee 0'$. We must argue that they are dense. Consider any p. We can clearly extend it to a q with $|q| \geq j$ by making $q(\langle e, x \rangle) = C_{|p|}(e)$ for $\langle e, x \rangle \in [|p|, j)$. So we may as well assume that $|p| \geq j$. If $\Phi_m^p(m) \downarrow$ then $p \in D_{m,j}$ and we are done. So suppose $\Phi_m^p(m) \uparrow$. If there is $q \supseteq p$ such that $\Phi_m^q(m) \downarrow$ and $(\forall e < m)(\forall \langle e, x \rangle \in [|p|, |q|)[C_{|p|}(x) = q(\langle e, x \rangle)$ or $\exists n \leq e\, (\Phi_n^p(n) \uparrow \ \& \ \Phi_n^q(n) \downarrow)]$, $q \leq_{\mathcal{P}} p$ by definition (because $\Phi_m^p(m) \uparrow$ while $\Phi_m^q(m) \downarrow$ so any violation of coding is allowed for $e \geq m$) and is in $D_{m,j}$. If there is no such q then $p \in D_{m,j}$ by definition.

Now we verify that $G = \cup p_s$ has the desired properties. By Theorem 5.21, $G \leq_T A$. To see that $C \leq_T G'$ consider any e. Let s be such that $(\forall i \leq e)(\Phi_i^G(i) \downarrow \Rightarrow \Phi_i^{p_s}(i) \downarrow \ \& \ i \in C \Rightarrow i \in C_{|p_s|})$. It is clear from the definition of $\leq_{\mathcal{P}}$ that for any $t > s$ and $\langle i, x \rangle \in [|p_s|, |p_t|)$ with $i \leq e$, $\langle i, x \rangle \in p_t \Leftrightarrow i \in C$. Thus $C(e) = \lim_t G(\langle e, t \rangle$ and so $C \leq_T G'$ by the Shoenfield limit lemma. For the other direction we want to compute $G'(e)$ recursively in C. (Of course, $A \leq_T C$ and so then is $\langle p_s \rangle$.) Suppose we have, by induction, computed an s as above for $e - 1$. We can now ask if $e \in C$. If so, we find a $u \geq t \geq s$ such that $e \in C_{|p_t|}$ and $p_u \in D_{e,|p_t|}$. If $\Phi_e^{p_u}(e) \downarrow$, then, of course, $e \in G'$. If $\Phi_e^{p_u}(e) \uparrow$ but $e \in G'$, then there would be a $v > u$ such that $\Phi_e^{p_v}(e) \downarrow$ and, of course, $p_v \leq_{\mathcal{P}} p_u$. This would contradict the fact that $p_u \in D_{e,|p_t|}$ by our choice of s and t and the definitions of $D_{e,|p_t|}$ and $\leq_{\mathcal{P}}$. □

Corollary 5.23: *(Shoenfield Jump Inversion Theorem) For all $C \geq$*

$0'$ there is $B < 0'$ such that $B' \equiv_T C$ if and only if C is r.e. in $0'$.

Proof: The "only if" direction is immediate. The "if" direction follows directly from the Theorem by taking $A = 0'$. □

For later applications we now strengthen the above jump inversion theorem to make $B <_T A$.

Theorem 5.24: *If $A \in \overline{GL_2}$, $C \geq_T A \vee 0'$, and C is r.e. in A, then there is $B <_T A$ such that $B' \equiv_T C$.*

Proof: In addition to the requirements of Theorem 5.22, we need to make sure that $\Phi_i^G \neq A$ for each i. To do this we modify the definition of extension to also allow violations of the coding requirements for e when we newly satisfy one of these diagonalization requirements for $i \leq e$. (As we did above for making $\Phi_i^G(i)\downarrow$.) We say $q \leq_1 p$ if

$$\langle e, x\rangle \in [|p|, |q|) \Rightarrow [C_{|p|}(x) = q(\langle e, x\rangle) \text{ or}$$

$$\exists n \leq e\,([\Phi_n^p(n)\uparrow \;\&\; \Phi_n^q(n)\downarrow] \text{ or } [\exists y \Phi_n^q(y)\downarrow \neq A(y) \;\&\; \neg\exists y \Phi_n^p(y)\downarrow \neq A(y)])\,.$$

Again $\leq_{\mathcal{P}}$ is defined as the transitive closure of this relation and it is recursive in $A \vee 0'$ as before. We then adjust the $D_{m,j}$ accordingly

$$D_{m,j} = \{p : |p| > j \;\&\; [\Phi_m^p(m)\downarrow \text{ or } (\forall q \supseteq p)(\Phi_m^q(m)\uparrow$$
$$\text{or } [(\exists e < m)(\exists \langle e, x\rangle \in [|p|, |q|)(C_{|p|}(e) \neq q(\langle e, x\rangle) \text{ but } \neg(\exists n \leq e)$$
$$([\Phi_n^p(n)\uparrow \;\&\; \Phi_n^q(n)\downarrow] \;\&\; \neg(\exists y)[\Phi_n^q(y)\downarrow \neq A(y) \;\&\; \neg\exists y \Phi_n^p(y)\downarrow \neq A(y)])]\}.$$

We also need dense sets that guarantee that $\Phi_e^G \neq A$:

$$D_i = \{p|(\exists x)(\Phi_i^p(x)\downarrow \neq A(x) \text{ or}$$
$$(\forall q_0, q_1 \supseteq p)(\forall x < |q_0|, |q_1|)[\neg(\Phi_i^{q_0}(x)\downarrow \neq \Phi_i^{q_1}(x)\downarrow) \text{ or}$$
$$((\exists e < i)(\exists \langle e, x\rangle \in [|p|, |q|)(\exists j \in \{0,1\})[(C_{|p|}(e) \neq q_i(\langle e, x\rangle) \text{ but } \neg(\exists n \leq i)$$
$$([\Phi_n^p(n)\uparrow \;\&\; \Phi_n^q(n)\downarrow] \;\&\; \neg(\exists y)[\Phi_n^q(y)\downarrow \neq A(y) \;\&\; \neg\exists y \Phi_n^p(y)\downarrow \neq A(y)])]\}.$$

The proof now proceeds as in the previous Theorem. The arguments for all the verifications are now essentially the same as there and are left as an exercise. □

Exercise 5.25: Verify that the notion of forcing and classes of dense sets specified in the proof of Theorem 5.24 suffice to actually prove it.

Exercise 5.26: Prove that if A is r.e. and $C \geq_T 0'$ is r.e. in A then there is a $B \leq_T A$ such that $B' \equiv_T C$. Indeed we may also make $B <_T A$.

The next result says that every $\mathbf{a} \in \overline{\mathbf{GL}_2}$ is **RRE** (*relatively recursively enumerable*), i.e. there is a $\mathbf{b} < \mathbf{a}$ such that $\mathbf{a}$ is r.e. in $\mathbf{b}$ and a bit more.

Theorem 5.27: *If* $\mathbf{a} \in \overline{\mathbf{GL}_2}$ *then there is* $\mathbf{b} < \mathbf{a}$ *such that* $\mathbf{a}$ *is r.e. in* $\mathbf{b}$ *and* $\mathbf{a}$ *is in* $\overline{\mathbf{GL}_2}(\mathbf{b})$, *i.e.* $(\mathbf{a} \vee \mathbf{b}')' < \mathbf{a}''$.

Proof: Let $\mathbf{a} \in \overline{\mathbf{GL}_2}$. We'll use a notion of forcing $\mathcal{P}$ with conditions $p = \langle p_0, p_1, p_2 \rangle$, $p_i \in 2^{<\omega}$ such that

(1) $|p_0| = |p_1|$, $p_0(d_n) = A(n)$, $p_1(d_n) = 1 - A(n)$ where d_n is nth place where p_0, p_1 differ and
(2) $(\forall e < |p_0 + p_1|)(e \in p_0 \oplus p_1 \Leftrightarrow \exists x(\langle e, x \rangle \in p_2))$.

As expected, our generic set $G_0 \oplus G_1 \oplus G_2$ is given by $V(p) = p_0 \oplus p_1 \oplus p_2$. The idea here is that if we can force p_0, p_1 to differ at infinitely many places while still making our generic sequence recursive in A, the first clause in the definition of $\leq_{\mathcal{P}}$ guarantees that $G_0 \oplus G_1 \equiv_T A$. The second clause works towards making $G_0 \oplus G_1$ r.e. in G_2 with the intention being that $\deg(G_2) = \mathbf{g}_2$ is to be the $\mathbf{b}$ required by the theorem. Extension in the notion of forcing is defined in the simplest way as $q \leq_{\mathcal{P}} p \Leftrightarrow q_i \supseteq p_i$ but note that this only applies to p and q in $\mathcal{P}$ and not all q with $q_i \supseteq p_i$ are in $\mathcal{P}$ even if $p \in \mathcal{P}$. The notion of forcing is clearly recursive in A.

We now define the dense sets needed to satisfy the requirements of the Theorem. We begin with $D_{2n} = \{p : p_0, p_1 \text{ differ at at least } n \text{ points}\}$. These sets are clearly recursive in A. We argue that these are dense by induction on n. Suppose D_{2n} is dense. To show that D_{2n+2} is dense, it suffices, for any given $p \in D_{2n} - D_{2n+2}$, to find a $q \leq_{\mathcal{P}} p$ in D_{2n+2}. Let $q_0 = p_0\hat{\ }A(n)$, $q_1 = p_1\hat{\ }(1 - A(n))$. Choose $i \in \{0, 1\}$ such that $q_i(|p_0|) = 1$. Define $q_2 \supseteq p_2$ by choosing x large and setting $q_2(\langle 2|p_0| + i, x \rangle) = 1$ and $q_2(z) = 0$ for all $z \notin \operatorname{dom}(p_2)$ and less than $\langle 2|p_0| + i, x \rangle$. Now $q = \langle q_0, q_1, q_2 \rangle$ satisfies the requirements to be a condition in P. It obviously extends p and is in D_{2n+2}.

For any generic recursive in A which meets all the D_{2n}, $G_0 \oplus G_1 \equiv_T A$ and $G_0 \oplus G_1$ is r.e. in G_2.

We also want dense sets similar in flavor to those of the previous theorems to force the jump of G_2 to make $(\mathbf{a} \vee \mathbf{g}_2')' < \mathbf{a}''$. Let

$$D_{2n+1} = \{p : \Phi_n^{p_2}(n) \downarrow \text{ or } (\forall \sigma \supseteq p_2)$$
$$(\Phi_n^{\sigma}(n) \uparrow \text{ or } (\exists \langle e, x \rangle \in \sigma)((p_0 \oplus p_1)(e) = 0)\}.$$

For $p \in P$, membership in D_{2n+1} is a $0'$ question and so these sets are recursive in $A \vee 0'$. We want to prove that they are dense. Suppose have a $p \in P$ and so we want a $q \leq_{\mathcal{P}} p$ with $q \in D_{2n+1}$. We may suppose that $\Phi_n^{p_2}(n) \uparrow$ and that the second clause fails for p as otherwise we would already be done. Thus we have a $\sigma \supseteq p_2$ such that $\Phi_n^{\sigma}(n) \downarrow$ but $\neg(\exists\langle e, x\rangle \in \sigma)((p_0 \oplus p_1)(e) = 0)$. We claim that there is a $q \leq_{\mathcal{P}} p$ such that $q_2 \supseteq \sigma$ and so $\Phi_n^{q_2}(n) \downarrow$ and $q \in D_{2n+1}$ as required. The only issue is that there may be some $\langle j, y\rangle \in \sigma$ with $j > |p_0 \oplus p_1|$. If so, we must define q_0 and q_1 accordingly, i.e. $j \in q_0 \oplus q_1$. So if j is even, we want $\frac{j}{2} \in q_0$; if it is odd, $\frac{j-1}{2} \in q_1$. We now define q_0, q_1 at the appropriate element ($\frac{j}{2}$ or $\frac{j-1}{2}$) to both be 1. Elsewhere we let both q_0 and q_1 be 0. Thus we have not added any points at which q_0 and q_1 differ beyond those in p_0, p_1). Now we extend σ to q_2 by adding $\langle e, y\rangle$ for some large y if $(q_0 \oplus q_1)(e) = 1$ and $e \geq |p_0 \oplus p_1|$ and wherever not yet defined we let $q_2(z) = 0$. Thus $q \in P$ and is the desired extension of p in D_{2n+1} as $\Phi_n^{q_2}(n) = \Phi_n^{\sigma}(n) \downarrow$.

We now let $\langle p_s\rangle \leq_T A$ be a generic sequence meeting every D_n as given by Theorem 5.21. We have already seen that $G_0 \oplus G_1 \equiv_T A$ and it is r.e. in $G_2 \leq_T A$. If we can show that $(A \oplus G_2')' <_T A''$ then we will be done as this clearly implies that $G_2 <_T A$. We first claim that $G_2' \leq_T A \vee 0'$. To see if $n \in G_2'$, recursively in $A \vee 0'$ find an s such that $p_s \in D_{2n+1}$. Then we claim that $n \in G_2' \Leftrightarrow \Phi_n^{p_{s,2}}(n) \downarrow$. If $\Phi_n^{p_2}(n) \downarrow$, then we are done. If not, then $(\forall \sigma \supseteq p_{s,2})\,(\Phi_n^{\sigma}(n) \uparrow$ or $(\exists\langle e, x\rangle \in \sigma)((p_0 \oplus p_1)(e) = 0))$ and by definition of membership and extension in $\mathcal{P}$, $\Phi_n^{p_{t,2}}(n) \uparrow$ for every $p_{t,2}$ for $t \geq s$. Thus $\Phi_n^{G_2}(n) \uparrow$ as desired. As $G_2' \leq_T A \vee 0'$, $(A \oplus G_2') = A \vee 0'$ and so as $A \notin GL_2$, $(A \oplus G_2')' = (A \vee 0')' <_T A''$ as required. □

Exercise 5.28: If $A >_T 0$ is r.e. and $C \geq_T 0'$ is r.e. in A then there is a $B \leq_T A$ such that $B' \equiv_T C$. Indeed we may also make $B <_T A$. Hint: Build β_s finite extensions that obey a coding rule for columns for $e \leq c(s) \leq s$ (so that we can enumerate C recursively in A) except that we can violate this rule so as to force the jump as above; search below $m_A(s+1)$ for extensions forcing the jump for $e \leq s$ that obey rule. Also search for extensions with Φ_e giving different answers and allow violations in columns $> e$ when we satisfy this requirement by choosing one that gives an answer other than A.

We can now deduce a result that plays a major role in our analysis of definability in $\mathcal{D}(\leq\mathbf{0}')$ (and many other results).

Theorem 5.29: *If* $\mathbf{b} <_T \mathbf{a}$ *and* $\mathbf{a} \in \overline{GL}_2(\mathbf{b})$ *and* $\mathcal{I}$ *is a* Σ_3^B *ideal in* $\mathcal{D}(\leq\mathbf{b})$

then there is an exact pair for $\mathcal{I}$ below $\mathbf{a}$.

Proof: By Theorem 5.27 (relativized to $\mathbf{b}$) there is a $\mathbf{c}$ such that $\mathbf{b} \leq \mathbf{c} < \mathbf{a}$ and $\mathbf{a}$ is r.e. in $\mathbf{c}$. So $\mathcal{I}$ is also Σ_3^C. Now, by Theorem 5.12, we have the desired exact pair. □

Theorem 5.30: *If $A \in \mathbf{a} \in \overline{\mathbf{GL}}_2$ and $S \in \Sigma_3^A$ then there is an embedding of a nice effective successor model (with the appropriate partial lattice structure) in the degrees below* $\deg(A)$ *and an exact pair $\mathbf{x}, \mathbf{y} \leq \mathbf{a}$ for the ideal generated by the $\mathbf{d}_n$ with $n \in S$. (Remember that the $\mathbf{d}_n$ are the degrees representing $n \in \mathbb{N}$ in the effective successor model.)*

Proof: Given $A \in \overline{GL}_2$ and $S \in \Sigma_3^A$, Theorem 5.27 gives us a $B < A$ such that A is r.e. in B and A is $\overline{GL}_2(B)$. Since $A' \geq A \vee 0'$ and is r.e. in it, Theorem 5.22 relativized to B gives us a $\hat{B} < A$ (with $B \leq_T \hat{B}$) such that $\hat{B}' \equiv A'$ and so $\Sigma_3^{\hat{B}} = \Sigma_3^A$, Moreover, A is r.e. in $\hat{B}$ because it was r.e. in $B \leq_T \hat{B}$. The result now follows by using Theorem 5.7 and Exercise 3.53 to embed an effective successor model between $\hat{B}$ and A and then Theorem 5.12 to pick out the ideal generated by the associated $\mathbf{d}_n$ for $n \in S$ as the set $\{e | \exists n(\Phi_e^{\hat{B}} \in \mathbf{d}_n)\}$ is itself $\Sigma_3^{\hat{B}} = \Sigma_3^A$ so is $\{e | (\exists n \in S)(\Phi_e^A \in \mathbf{d}_n)\}$. □

Exercise 5.31: Prove that every degree has a $\mathbf{GL}_2$ degree below it.

Exercise 5.32: Prove that every recursive lattice $\mathcal{L}$ with 0 and 1 can be embedded in $\mathcal{D}(\leq \mathbf{a})$ preserving 0 and 1 for any $\mathbf{a} \in \overline{\mathbf{GL}_2}$.

Notes: Theorem 5.14 is due to Martin [1966]. Its very useful consequence, Proposition 5.15 is from Jockusch and Posner [1978] which also contains a version of Theorem 5.16 for Cohen forcing, Exercises 5.17 and 5.31 as well as Theorem 5.22. The version given here of Theorem 5.16 and the more general Theorem 5.21 as well as Theorem 5.27 come from Cai and Shore [2012]. Corollary 5.23 was originally proved in Shoenfield [1959]. The original direct proof of (a stronger version of) Theorem 5.30 is in Shore [2007]. Remark 5.20 follows, for example, from Slaman and Steel [1989, Theorem 3.1] or Cooper [1989]. Theorem 5.19 is from Jockusch and Posner [1978].

5.4. *Definability and biinterpretability in $\mathcal{D}(\leq \mathbf{0}')$*

We already know that the theory of $\mathcal{D}(\leq \mathbf{0}')$ is (recursively) equivalent to true first order arithmetic and so as complicated as possible. We now want attack the problem of determining which subsets of, and relations on, $\mathcal{D}(\leq\mathbf{0}')$ are definable in the structure. The interpretation of $\mathcal{D}(\leq\mathbf{0}')$ in $\mathbb{N}$ gives a necessary condition. Only subsets and relations definable in arithmetic can possibly be definable in $\mathcal{D}(\leq\mathbf{0}')$. Our goal is to prove that, if they are also invariant under the double jump, then the are, in fact, definable in $\mathcal{D}(\leq\mathbf{0}')$.

Definition 5.33: A relation $R(x_1,\ldots,x_n)$ on degrees is invariant under the double jump if, for all degrees $\mathbf{x}_1,\ldots,\mathbf{x}_n$ and $\mathbf{y}_1,\ldots,\mathbf{y}_n$ such that $\mathbf{x}_i''=\mathbf{y}_i''$ for all $i\leq n$, $R(\mathbf{x}_1,\ldots,\mathbf{x}_n)\Leftrightarrow R(\mathbf{y}_1,\ldots,\mathbf{y}_n)$.

We begin with the subsets of $\mathcal{D}(\leq\mathbf{0}')$ and, in particular, with the basic question of definably determining the double jump of a degree $\mathbf{a}\leq\mathbf{0}'$. (This would actually suffice to show that all subsets of $\mathcal{D}(\leq\mathbf{0}')$ invariant under double jump and definable in arithmetic are definable in $\mathcal{D}(\leq\mathbf{0}')$ but as we prove more later we omit this argument.) The crucial point is that the sets we can code below an r.e. or $\overline{\mathbf{GL}_2}$ degree $\mathbf{a}$ are precisely the ones Σ_3^A. We use this to determine $\mathbf{a}''$ via the following characterization of the double jump.

Proposition 5.34: *For any sets A and B, $A''\equiv_T B''$ if and only if $\Sigma_3^A=\Sigma_3^B$. Indeed, for any $n\geq 1$, $A^{(n)}\equiv_T B^{(n)}$ if and only if $\Sigma_{n+1}^A=\Sigma_{n+1}^B$.*

Proof: The hierarchy theorem 1.10 says that, for any set X and $n\geq 1$, $\Sigma_{n+1}^X=\Sigma_1^{X^{(n)}}$. On the other hand, for any Z and W, $\Sigma_1^Z=\Sigma_1^W$ iff $Z\equiv_T W$ since the equality implies that both Z and $\bar{Z}$ (W and $\bar{W}$) are Σ_1, i.e. r.e., in W (Z) and so each is recursive in the other. Thus if $\Sigma_{n+1}^A=\Sigma_{n+1}^B$ then $\Sigma_1^{A^{(n)}}=\Sigma_1^{B^{(n)}}$ and so $A^{(n)}\equiv_T B^{(n)}$ as required. □

Theorem 5.35: *The set $\mathbf{L}_2=\{\mathbf{x}\leq\mathbf{0}'|\mathbf{x}''=\mathbf{0}''\}$ is definable in $\mathcal{D}(\leq\mathbf{0}')$.*

Proof: Our analysis of coding in models of arithmetic in Proposition 3.59 and preceding Theorem 4.10 (which is really part of the proof of that theorem), shows that we have a way to, definably in $\mathcal{D}(\leq\mathbf{0}')$, pick out, via correctness conditions, parameters $\bar{\mathbf{p}}$ that define structures $\mathcal{M}(\bar{\mathbf{p}})$ isomorphic to $\mathbb{N}$. (The crucial point here is Theorem 5.12 which says that there is an exact pair for the $\Sigma_3^{\bar{\mathbf{p}}_0}$ ideal generated by the standard part of the model below $\mathbf{0}'$ as it is r.e. in and strictly above $\bar{\mathbf{p}}_0$.) Also note that, by Proposition 3.59, any set S coded in $\mathcal{M}(\bar{\mathbf{p}})$ by a pair $\mathbf{g}_0,\mathbf{g}_1$ and a coding formula $\varphi_S(x,\bar{\mathbf{p}})$ is Σ_3^A as long as the parameters $\bar{\mathbf{q}}$ for the nice effective successor structure determining the domain of the model and $\mathbf{g}_0, \mathbf{g}_1$ are recursive in A.

We now claim that $\mathbf{x} \in \mathbf{L}_2$ if and only for any such $\bar{\mathbf{q}}, \mathbf{g}_0, \mathbf{g}_1 \leq_T \mathbf{x}$ the set S coded by $\mathbf{g}_0, \mathbf{g}_1$ is Σ_3. Moreover, this property is definable in $\mathcal{D}(\leq \mathbf{0}')$ and so proves the Theorem.

First suppose that $\mathbf{x} \in \mathbf{L}_2$. Then our initial remarks show that $S \in \Sigma_3^X$ for any $X \in \mathbf{x}$. As $X'' \equiv_T 0''$, $\Sigma_3^X = \Sigma_3$ by Proposition 5.34. Next, if $\mathbf{x} \notin \mathbf{L}_2$, then by Exercise 3.53 and Theorem 5.16 there are parameters $\bar{\mathbf{q}}$ defining a nice effective successor model with join $\mathbf{c} < \mathbf{x}$ with $\mathbf{c}' = \mathbf{0}'$. By Theorem 4.9, we can extend these parameters to ones $\bar{\mathbf{p}}$ defining a standard model of arithmetic which, of course, satisfies the definable properties guaranteeing that it is such a model. Now, by Theorem 5.29, for any $S \in \Sigma_3^X$ there are $\mathbf{g}_0, \mathbf{g}_1 \leq_T \mathbf{x}$ which code S in this model. Since $\mathbf{x}'' > \mathbf{0}''$ there is an $S \in \Sigma_3^X - \Sigma_3$ again by Proposition 5.34 and so a code for such an S below $\mathbf{x}$ as required.

Finally, note that, as we are working in definable standard models of arithmetic, we can definably say that a set is Σ_3 simply by using the translation into our degree structure of the corresponding sentence of arithmetic. □

Theorem 5.36: *For every $\mathbf{h} \geq \mathbf{0}''$ which is r.e. in $\mathbf{0}''$, the set $\{\mathbf{x} \leq \mathbf{0}' | \mathbf{x}'' = \mathbf{h}\}$ is definable in $\mathcal{D}(\leq \mathbf{0}')$.*

Proof: The previous theorem handles the case that $\mathbf{h} = \mathbf{0}''$. For $\mathbf{h} > \mathbf{0}''$ Let $E \in \mathbf{e} \in [\mathbf{0}', \mathbf{0}'']$ be such that $E' \in \mathbf{h}$. There is such an E by Corollary 5.23 and we can fix a definition of one in arithmetic. Consider the formula which says that for any $\mathbf{q}, \mathbf{g}_0, \mathbf{g}_1 < \mathbf{x}$ and $\bar{\mathbf{p}}$ which define a standard model of arithmetic and a set S coded in the model as in the proof of the Theorem, $S \in \Sigma_2^E$ and for any set $\hat{S} \in \Sigma_2^E$ (again as given by a definition in arithmetic) there are such $\mathbf{q}, \mathbf{g}_0, \mathbf{g}_1 < \mathbf{x}$ and $\bar{\mathbf{p}}$ defining $\hat{S}$. Proposition 5.34 and calculations already described now show that this guarantees that $\Sigma_2^{X'} = \Sigma_3^X = \Sigma_2^E$ and so $\mathbf{x}'' = \mathbf{e}' = \mathbf{h}$ as required. □

Corollary 5.37: *The jump classes $\mathbf{L}_n$ ($\mathbf{a}^{(n)} = \mathbf{0}^{(n)}$) and $\mathbf{H}_n$ ($\mathbf{a}^{(n)} = \mathbf{0}^{(n+1)}$) are definable in $\mathcal{D}(\leq \mathbf{0}')$ for $n \geq 2$.*

Proof: In the proof of Theorem 5.36, require instead of $E' \in \mathbf{h}$ that $E^{(n-1)} \equiv_T 0^{(n)}$ for $\mathbf{L}_n$ and $E^{(n-1)} \equiv_T 0^{(n+1)}$ for $\mathbf{H}_n$. □

By a separate additional argument that requires results beyond the scope of these lectures we can also get the definability of $\mathbf{H}_1$. While we could make such an argument at this point it will be easier later. We do so in Corollary 5.43. The definability of $\mathbf{L}_1$ in $\mathcal{D}(\leq\mathbf{0}')$ is an important open problem.

If we now wish to deal with arbitrary relations on $\mathcal{D}(\leq\mathbf{0}')$ rather than simply subsets, we are faced with the problem that our analysis so far has, for each degree $\mathbf{a}$, produced various models of arithmetic in which we code the sets Σ_3^A. To discuss even binary relations we must have a way to analyze any $\mathbf{a}$ and $\mathbf{b}$ (or equivalently the sets coded below them as long as we are only working up to invariance under the double jump) in a single model (perhaps with additional correctness conditions). The basic formulation of this issue is given by asking about the biinterpretability of the structure (here $\mathcal{D}(\leq\mathbf{0}')$) with arithmetic (here first order). A similar notion applies to other structures (such as the r.e. degrees, $\mathcal{R}$) still with first order arithmetic and to ones such as $\mathcal{D}$ but for second order arithmetic.

Definition 5.38: A degree structure $\mathcal{S}$ is *biinterpretable with true first (second) order arithmetic* if it is interpretable in first (second) order arithmetic and we have formulas in parameters $\bar{\mathbf{p}}$ (including a correctness condition) as specified in Sec. 4.1 (and, for the second order case, a formula $\varphi_S(x, \bar{y})$ which defines sets (coded) in the model given by $\bar{\mathbf{p}}$) as described there which provide an interpretation of true arithmetic in $\mathcal{S}$ (i.e. the models $\mathcal{M}(\bar{\mathbf{p}})$ satisfying the correctness condition are all standard). For second order arithmetic, we also require that the sets defined by $\varphi_S(x, \bar{y})$ as $\bar{y}$ ranges over all parameters in $\mathcal{S}$ are all subsets of $\mathbb{N}$.

Moreover, for both first and second order arithmetic, there is an additional formula $\varphi_R(x, \bar{y}, \bar{\mathbf{p}})$ such that $\mathcal{S} \models \forall x \exists \bar{y} \varphi_R(x, \bar{y}, \bar{\mathbf{p}})$ and for every $\mathbf{a}, \bar{\mathbf{g}} \in \mathcal{S}$, $\mathcal{S} \models \varphi_R(\mathbf{a}, \bar{\mathbf{g}}, \bar{\mathbf{p}})$ if and only if the set $\{n | \varphi_S(\mathbf{d}_n, \bar{\mathbf{g}}, \bar{\mathbf{p}})\}$ (where $\mathbf{d}_n$ is the nth element of the model $\mathcal{M}(\bar{\mathbf{p}})$ coded by the parameters $\bar{\mathbf{p}}$) is of degree $\mathbf{a}$. These last conditions then say that the set coded in $\mathcal{M}(\bar{\mathbf{p}})$ by $\bar{\mathbf{g}}$ is of degree $\mathbf{a}$ and that all degrees $\mathbf{a}$ in $\mathcal{S}$ have codes $\bar{\mathbf{g}}$ for a set of degree $\mathbf{a}$.

We say that $\mathcal{S}$ is *biinterpretable with true first or second order arithmetic up to double jump* if we weaken the second condition on φ_R so that for every $\mathbf{a}, \bar{\mathbf{g}} \in S$, $\mathcal{S} \models \varphi_R(\mathbf{a}, \bar{\mathbf{g}}, \bar{\mathbf{p}})$ if and only if the set $\{n | \varphi_S(\mathbf{d}_n, \bar{\mathbf{g}}, \bar{\mathbf{p}})\}$ has the same double jump as $\mathbf{a}$.

It is not hard to see that, if a degree structure $\mathcal{S}$ is biinterpretable with first or second order arithmetic, then we know all there is to know about definability in, and automorphisms of, $\mathcal{S}$.

Theorem 5.39: *If a degree structure $\mathcal{S}$ is biinterpretable with first or second order arithmetic then it is rigid, i.e. it has no automorphisms other than the identity, and a relation on $\mathcal{S}$ is definable in $\mathcal{S}$ if and only if it is definable in first or second order arithmetic, respectively.*

Proof: We first prove rigidity. Let $\bar{\mathbf{p}}$ satisfy all the formulas required for it to determine a standard model of arithmetic via the given formulas. Consider any $\mathbf{a} \in \mathcal{S}$ with some $\bar{\mathbf{g}}$ such that $\mathcal{S} \vDash \varphi_R(\mathbf{a}, \bar{\mathbf{g}}, \bar{\mathbf{p}})$ and any automorphism Ψ of $\mathcal{S}$. The image $\Psi(\bar{\mathbf{p}}) = \bar{\mathbf{r}}$ satisfies all the same formulas as $\bar{\mathbf{p}}$ and so also defines a standard model of arithmetic. The image $\bar{\mathbf{h}}$ of $\bar{\mathbf{g}}$ under Ψ also determines a subset of this model via φ_S and it must be the "same" subset in the sense that they correspond to the same subset of $\mathbb{N}$ via the isomorphisms among $\mathcal{M}(\bar{\mathbf{p}})$, $\mathcal{M}(\bar{\mathbf{r}})$ and $\mathbb{N}$. Of course, $\varphi_R(\mathbf{b}, \bar{\mathbf{h}}, \bar{\mathbf{r}})$ (where $\mathbf{b} = \Psi(\mathbf{a})$) is also true in $\mathcal{S}$ since Ψ is an automorphism. Our definition of biinterpretability now says that $\mathbf{a} = \mathbf{b}$ as required for rigidity.

Now consider any relation $Q(\bar{x})$ on $\mathcal{S}$. By the assumption that $\mathcal{S}$ is interpretable in first or second order arithmetic, we know that Q is definable in those structures. For the other direction, suppose Q is definable by a formula Θ of first or second order arithmetic. If this is first order arithmetic then we expanded it by a sequence $\bar{X}$ of second order parameters (of the same length n as $\bar{x}$) whose intended interpretations are some subsets of the model. If it is second order arithmetic then we simply assume that the formula already contains a sequence $\bar{X}$ of free second order variables (of the same length as $\bar{x}$). In any case, Θ defines the property that the sequence of the degrees of X satisfies Q.) Q is then defined in $\mathcal{S}$ by the formula $\Psi(\bar{z}) \equiv \exists \bar{p}, \bar{g}_0 \ldots \exists \bar{g}_{n-1}(\varphi_c(\bar{p})\ \&\ \bigwedge_{i<n} \varphi_R(z_i, \bar{g}_i, \bar{p}) \rightarrow \Theta^T(\bar{g}_i, \bar{p}))$ where T is the translation of formulas of second order arithmetic given in Sec. 4.1. Here our correctness condition φ_c guarantees that the model $\mathcal{M}(\bar{p})$ is standard and we also assume that the requirements of the definition of biinterpretability are satisfied. So the translation of Θ asserts (because of the properties of φ_R) that a sequence of sets of degree z_i satisfy Θ (in $\mathbb{N}$), i.e. Q holds of $\bar{z}$. □

Our goal now is to prove that $\mathcal{D}(\leq \mathbf{0}')$ is biinterpretable with arithmetic up to double jump and so every relation on it invariant under the double jump is definable in it if and only if it is definable in first order arithmetic.

Theorem 5.40: *$\mathcal{D}(\leq \mathbf{0}')$ is biinterpretable with arithmetic up to double jump.*

Theorem 5.35 and Theorem 5.36 show that we can define the double jump classes of degrees $\mathbf{a}$ in $\mathcal{D}(\leq\mathbf{0}')$ by talking about the sets that are coded (by our usual formula $\varphi_S(x, \bar{\mathbf{g}})$) in standard models $\mathcal{M}(\bar{\mathbf{p}})$ of arithmetic with $\bar{\mathbf{q}}, \bar{\mathbf{g}}$ below $\mathbf{a}$ as in the proof of Theorem 5.35. The point here is that these sets determine Σ_3^A and so $\mathbf{a}''$ by Proposition 5.34. If we wish to define the relations needed for biinterpretability up to double jump, we need to be able to talk about the sets that are Σ_3^A for an arbitrary degree $\mathbf{a}$ simultaneously in a single model. Our plan is to provide a scheme defining isomorphisms between two arbitrary standard models satisfying some additional correctness condition. Such isomorphisms would allow us to definably transfer (codes for) sets in different models to ones for the same sets in a single model and so define the required relation φ_R. We begin with a lemma that is used to build such isomorphisms by interpolating a sequence of additional models between the two given ones and isomorphisms between each successive pair of models.

Lemma 5.41: *If $\mathbf{c} \leq \mathbf{0}'$, $\mathbf{c} \in \bar{\mathbf{L}}_2, \mathbf{a}_0, \mathbf{a}_1 \in \mathbf{L}_1$ and $\mathcal{P}$ is a recursive notion of forcing, then there is a $G \leq_T C$ which is 1-generic for $\mathcal{P}$ and such that $A_0 \oplus G$ and $A_1 \oplus G$ are both low.*

Proof: Let $D_{n,2}$ be the usual dense sets for making G 1-generic for $\mathcal{P}$. They, and the density function for them, are uniformly recursive in $0'$. Now consider, for $i \in \{0,1\}$, the sets $D_{n,i} = \{p | \Phi_n^{A_i \oplus V(p)}(n) \downarrow \text{ or } (\forall q \leq p)\Phi_n^{A_i \oplus V(q)}(n) \uparrow\}$. As the A_i are low, these sets and their density functions are also uniformly recursive in $0'$. Thus, by Theorem 5.21, there is a 1-generic sequence $\langle p_k \rangle$ and an associated generic set G both recursive in C meeting all these dense sets. Any such G clearly has all the properties required in the theorem. (Follow, for example, the proof of Proposition 3.32 using these $D_{e,i}$ in place of $D_{1,e}$.) □

Proof: (**of Theorem 5.40**) In addition to the previous correctness conditions for our standard models $\mathcal{M}(\bar{\mathbf{p}})$ we require for the rest of this subsection that $\mathbf{p}_0$, the first of the parameters $\bar{\mathbf{p}}$, which bounds the parameters $\bar{\mathbf{q}}$ defining the nice effective successor structure providing the domain $\mathbf{d}_n$ of the model, is in $\bar{\mathbf{L}}_2$. (This condition is definable by Theorem 5.35.) Given two such models $\mathcal{M}(\bar{\mathbf{p}}_0)$ and $\mathcal{M}(\bar{\mathbf{p}}_4)$ we want to show that there are additional models $\mathcal{M}(\bar{\mathbf{p}}_k)$ for $k \in \{1,2,3\}$ and uniformly definable isomorphisms between the domains of these models taking $\mathbf{d}_{i,n}$ to $\mathbf{d}_{1+1,n}$ for $i < 4$. (Given parameters $\bar{\mathbf{p}}_k$ defining a model $\mathcal{M}(\bar{\mathbf{p}}_k)$ we write $\mathbf{d}_{k,n}$ for the degree representing the nth element of this model. Similarly, we write $\bar{\mathbf{p}}_{k,0}$ for the first element of $\bar{\mathbf{p}}_k$ and $\bar{\mathbf{q}}_k$ for the parameters in $\bar{\mathbf{p}}_k$ determining the effective successor structure which provides the domain of $\mathcal{M}(\bar{\mathbf{p}}_k)$.) Thus (as we explain below) we produce a single formula $\theta(x, y, \bar{z}, \bar{z}')$ which uniformly defines isomorphisms between any two of our standard models $\mathcal{M}(\bar{\mathbf{p}}_0)$ and $\mathcal{M}(\bar{\mathbf{p}}_4)$ (with $\bar{z}$ and $\bar{z}'$ replaced by $\bar{\mathbf{p}}_0$ and $\bar{\mathbf{p}}_4$).

We begin by choosing $\bar{\mathbf{q}}_1 < \mathbf{0}'$ as given by a 1-generic over $\mathbf{p}_{0,0}$ sequence and function for the recursive notion of forcing (Exercise 3.53) that embeds a nice effective successor model with $\bar{\mathbf{q}}_{1,0}$, the first element of $\bar{\mathbf{q}}_1$, being the bound on all the other required parameters. As $\mathbf{p}_{0,0} \in \mathbf{L}_2$, $\mathbf{0}'$ is $\bar{\mathbf{L}}_2(\mathbf{p}_{0,0})$ and so such $\bar{\mathbf{q}}_1$ exists by Theorem 5.16 (relativized to $\mathbf{p}_{0,0}$). Note that $\bar{\mathbf{q}}_1$ (and so $\bar{\mathbf{q}}_{1,0}$) is in $\mathbf{L}_1$ by Proposition 3.32 as it is associated with a 1-generic sequence recursive in $\mathbf{0}'$. We may now extend $\bar{\mathbf{q}}_1$ to $\bar{\mathbf{p}}_1$ defining a standard model $\mathcal{M}(\bar{\mathbf{p}}_1)$ by Exercise 4.8 and Theorem 5.16 as $\mathbf{0}'$ is $\overline{\mathbf{GL}}_2(\bar{\mathbf{q}}_1)$. Similarly, we see that there are $\bar{\mathbf{q}}_3$ and $\bar{\mathbf{p}}_3$ bearing the same relation to $\mathcal{M}(\mathbf{p}_4)$ as $\bar{\mathbf{q}}_1$ and $\bar{\mathbf{p}}_1$ do to $\mathcal{M}(\mathbf{p}_0)$. Now as $\bar{\mathbf{q}}_{1,0}$ and $\bar{\mathbf{q}}_{3,0}$ are both low we may apply Lemma 5.41 to the forcing of Exercise 3.53 to get $\bar{\mathbf{q}}_2 < \mathbf{0}'$ (again as $\mathbf{0}' \in \bar{\mathbf{L}}_2(\bar{\mathbf{q}}_{1,0}), \bar{\mathbf{L}}_2(\bar{\mathbf{q}}_{3,0})$) such that both $\bar{\mathbf{q}}_{1,0} \oplus \bar{\mathbf{q}}_{2,0}$ and $\bar{\mathbf{q}}_{2,0} \oplus \bar{\mathbf{q}}_{3,0}$ are in $\mathbf{L}_1$ and then extend $\bar{\mathbf{q}}_2$ to $\bar{\mathbf{p}}_2$ defining $\mathcal{M}(\bar{\mathbf{p}}_2)$ as we did for $\bar{\mathbf{q}}_1$.

We now apply Exercise 4.8 and Theorem 5.16 to get the desired schemes defining our desired isomorphisms: Given any $n \in \mathbb{N}$ and $i < 4$, consider the finite sequences of degrees $\langle \mathbf{d}_{i,0}, \dots, \mathbf{d}_{i,n} \rangle$ and $\langle \mathbf{d}_{i+1,0}, \dots, \mathbf{d}_{i+1,n} \rangle$. We want to show that there are parameters $\bar{\mathbf{r}}_i < \mathbf{0}'$ such that the formula $\varphi_2(x, y, \bar{\mathbf{r}}_i)$ (where $\varphi_2(x, y, \bar{z})$ ranges over binary relations as $\bar{z}$ varies as in Theorem 4.3) defines an isomorphism taking $\mathbf{d}_{i,k}$ to $\mathbf{d}_{i+1,k}$ for each $k \leq n$. By the results just cited it suffices to show that the $\bigoplus_{k<n} \mathbf{d}_{i,k} \oplus \bigoplus_{k<n} \mathbf{d}_{i+1,k}$ are in $\mathbf{L}_2$ for each $i < 4$. For $i = 0$, note that $\bar{\mathbf{q}}_1$ is associated with a 1-generic/$\mathbf{p}_{0.0}$ sequence which is recursive in $\mathbf{0}'$. Thus by Proposition 3.32 (suitably relativized) $(\bar{\mathbf{q}}_1 \oplus \mathbf{p}_{0,0})' = \mathbf{p}_{0,0}'$ and so $(\mathbf{q}_{1,0} \oplus \mathbf{p}_{0,0})' = \mathbf{p}_{0,0}'$. As $\mathbf{p}_{0,0} \in \mathbf{L}_2$, $\mathbf{0}'' = \bar{\mathbf{p}}_{0,0}'' = (\bar{\mathbf{q}}_{1,0} \oplus \bar{\mathbf{p}}_{0,0})''$ as required. The argument for $i = 3$ is similar. For the other pairs, we have already guaranteed that $\bar{\mathbf{q}}_{1,0} \oplus \bar{\mathbf{q}}_{2,0}$ and $\bar{\mathbf{q}}_{2,0} \oplus \bar{\mathbf{q}}_{3,0}$ are both $\mathbf{L}_1$.

We can now define the desired isomorphism $\theta(\mathbf{n}, \mathbf{m}, \bar{\mathbf{p}}_0, \bar{\mathbf{p}}_4)$ between $\mathcal{M}(\bar{\mathbf{p}}_0)$ and $\mathcal{M}(\bar{\mathbf{p}}_4)$. We say that an $\mathbf{n}$ in the domain of $\mathcal{M}(\bar{\mathbf{p}}_0)$ (i.e. $\varphi_D(\mathbf{n}, \bar{\mathbf{p}}_0)$) is taken to $\mathbf{m}$ in the domain of $\mathcal{M}(\bar{\mathbf{p}}_4)$ if and only if there are degrees $\bar{\mathbf{p}}_k$ for $k \in \{1, 2, 3\}$ defining models of arithmetic $\mathcal{M}(\bar{\mathbf{p}}_k)$ and ones $\bar{\mathbf{r}}_i$ for $i < 4$ as above such that each $\varphi_2(x, y, \bar{\mathbf{r}}_i)$ defines an isomorphism between initial segments of (the domains of) $\mathcal{M}(\bar{\mathbf{p}}_i)$ and $\mathcal{M}(\bar{\mathbf{p}}_{i+1})$ where the initial segment in $\mathcal{M}(\bar{\mathbf{p}}_0)$ is the one with largest element $\mathbf{n}$ and that in

$\mathcal{M}(\bar{\mathbf{p}}_1)$ has largest element $\mathbf{m}$. Clearly this can all be expressed using the formulas $\varphi_D(x, \bar{\mathbf{p}}_k)$ and $\varphi_<(x, y, \bar{\mathbf{p}}_k)$ defining the domains of $\mathcal{M}(\bar{\mathbf{p}}_k)$ and the orderings on them. Note that the definition of this isomorphism is uniform in $\bar{\mathbf{p}}_0$ and $\bar{\mathbf{p}}_4$ and that we have shown that for any $\bar{\mathbf{p}}_0$ and $\bar{\mathbf{p}}_4$ defining our standard models of arithmetic, there are parameters below $\mathbf{0}'$ defining all these isomorphisms. In other words, we have described the desired formula $\theta(x, \bar{y}, \bar{z}, \bar{z}')$.

We now wish to define the formula $\varphi_R(x, \bar{y}, \bar{\mathbf{p}}_0)$ required in the definition of biinterpretability up to double jump (for $\mathcal{M}(\bar{\mathbf{p}}_0)$ a model of arithmetic). (We have replaced $\bar{\mathbf{p}}$ in Definition 5.38 by $\bar{\mathbf{p}}_0$ to match our current notation.) First, φ_R says that, if $x \in \mathbf{L}_2$ (as defined by Theorem 5.35), then $\bar{y}$ defines (via our standard φ_S) the empty set in $\mathcal{M}(\bar{\mathbf{p}}_0)$. In addition, φ_R says that, if $x \notin \mathbf{L}_2$ and S is the set defined in $\mathcal{M}(\bar{\mathbf{p}}_0)$ by $\bar{y}$, then for every set $\hat{S} \in \Sigma_3^S$ (with $\hat{S}$ defined by other parameters $\bar{\mathbf{h}}$ in $\mathcal{M}(\bar{\mathbf{p}}_0)$ and $\hat{S} \in \Sigma_3^S$ expressed in the translation of arithmetic into $\mathcal{M}(\bar{\mathbf{p}}_0)$), there are $\bar{\mathbf{g}} < \mathbf{x}$ and $\bar{\mathbf{p}}_4$ with $\bar{\mathbf{p}}_{4,0} < \mathbf{x}$ such that $\bar{\mathbf{g}}$ codes a set $\hat{S}_4$ in $\mathcal{M}(\bar{\mathbf{p}}_4)$ and, for every $\mathbf{n}$ and $\mathbf{m}$, $\theta(\mathbf{n}, \mathbf{m}, \bar{\mathbf{p}}_0, \bar{\mathbf{p}}_4)$ implies that $\varphi_S(\mathbf{n}, \bar{\mathbf{h}}, \bar{\mathbf{p}}_0) \Leftrightarrow \varphi_S(\mathbf{m}, \bar{\mathbf{g}}, \bar{\mathbf{p}}_4)$, i.e. $\hat{S} = \hat{S}_4$. By all that we have done already, this guarantees that every $\hat{S} \in \Sigma_3^S$ is Σ_3^X. For the other direction, φ_R also says that if $\bar{\mathbf{g}} < \mathbf{x}$ and $\bar{\mathbf{p}}_4$ with $\bar{\mathbf{p}}_{4,0} < \mathbf{x}$ are such that $\bar{\mathbf{g}}$ codes a set $\hat{S}_4$ in $\mathcal{M}(\bar{\mathbf{p}}_4)$ then there is a set $\hat{S}$ (coded in $\mathcal{M}(\bar{\mathbf{p}}_0)$ by some $\bar{\mathbf{h}}$) which is Σ_3^S (as expressed in $\mathcal{M}(\bar{\mathbf{p}}_0)$) such that $\hat{S} = \hat{S}_4$ as expressed as above using θ. So again by what we have already done, this guarantees that every $\hat{S}_4 \in \Sigma_3^X$ is Σ_3^S. Thus, by Proposition 5.34, S has the same double jump as X as required. □

Theorem 5.42: *A relation on $\mathcal{D}(\leq \mathbf{0}')$ which is invariant under the double jump is definable in $\mathcal{D}(\leq \mathbf{0}')$ if and only if it is definable in true first order arithmetic.*

Proof: Follow the proof of Theorem 5.39 but use Theorem 5.40 in place of the assumption that the structure is biinterpretable with arithmetic. □

Corollary 5.43: $\mathbf{H}_1$ *is definable in* $\mathcal{D}(\leq \mathbf{0}')$.

Proof: This follows immediately from the Theorem and fact that $\mathbf{x} < \mathbf{0}'$ is in $\mathbf{H}_1$ if and only if $\mathcal{D}(\leq \mathbf{0}') \vDash \forall z \exists y \leq x (z'' = y'')$. This fact is proven for r.e. $\mathbf{x}$ in Nies, Shore and Slaman [1998, Theorem 2.21] but (as indicated there on p. 257) replacing the last use of the Robinson jump interpolation theorem in the proof by Theorem 5.22 provides a proof for $\mathcal{D}(\leq \mathbf{0}')$. □

The analogous theorems hold for both $\mathcal{D}$ and $\mathcal{R}$, i.e. they are biinterpretable with second or first order arithmetic, respectively, up to double jump. (Moreover, in $\mathcal{D}$ the jump is also definable.) Their definable relations which are invariant under the double jump are then characterized in the same way. Indeed, every jump ideal $\mathcal{I}$ of $\mathcal{D}$ (i.e. an ideal that is also closed under the jump operator) which contains $\mathbf{0}^{(\omega)}$ is biinterpretable with second order arithmetic up to double jump if one takes the second order structure to have sets precisely those with degrees in $\mathcal{I}$ and the jump is definable in $\mathcal{I}$ as well.

By more extensive uses of Theorem 5.21 we can prove our biinterpretability and so definability results for $\mathcal{D}(\leq \mathbf{x})$ for any $\mathbf{x} \leq \mathbf{0}'$ in $\bar{\mathbf{L}}_2$.

Exercise 5.44: For every $\mathbf{x} \leq \mathbf{0}'$ in $\bar{\mathbf{L}}_2$, $Th(\mathcal{D}(\leq\mathbf{x})$ is biinterpretable with true first order arithmetic and so its theory is 1-1 equivalent to that of true arithmetic. Moreover, for every $\mathbf{x} \leq \mathbf{0}'$ a relation on $\mathcal{D}(\leq \mathbf{x})$ invariant under double jump is definable in $\mathcal{D}(\leq \mathbf{x})$ if and only if it is definable in first order arithmetic. (For $\mathbf{x} \in \mathbf{L}_2$, this last result is trivial. Otherwise, it follows from biinterpretability as before.)

The *Biinterpretability Conjectures* for $\mathcal{D}(\leq\mathbf{0}')$, $\mathcal{R}$ and $\mathcal{D}$ assert that these structures are actually biinterpretable with first, first and second order arithmetic, respectively. As we have seen proofs of these conjectures would show that the structures are rigid and would completely characterize their definable relations. These are the major open problems of degree theory.

Notes: The definitions of biinterpretability for different degree structures and the associated conjectures are due to Harrington and Slaman and Woodin (see Slaman [1991] and [2008]). Theorems 5.35 and 5.36 are originally due to Shore [1988] but for triple jump in place of double jump. The improvement of one jump is essentially an application of Proposition 5.34 as pointed out in Nies, Shore and Slaman [1998] where Corollary 5.43 also appears. Slaman and Woodin also proved Theorem 5.39 (again see Slaman [1991] and [2008]). Plans for proving Theorem 5.40 were proposed in both Shore [1988] and more concretely in Nies, Shore and Slaman [1998] but neither actually provided the definitions of the required comparison maps nor the proofs that they exist as we have done here. Thus Theorems 5.40, 5.42 and the improvement to initial segments of $\mathcal{D}(\leq\mathbf{0}')$ bounded by any $\mathbf{x} \in \bar{\mathbf{L}}_2$ of Exercise 5.44 are new. A proof of Exercise 5.44 appears in Shore [2014]. Biinterpretability up to double jump for the r.e. degree $\mathcal{R}$ is proven in Nies, Shore and Slaman [1998]. Slaman and Woodin (see Slaman [1991] and [2008]) proved it for $\mathcal{D}$. A very different proof that also gives the results described for jump ideals containing $\mathbf{0}^{(\omega)}$ is in Shore [2007]. The definability of the jump is proven in Shore and Slaman [1999] based on the results of Slaman and Woodin. This reliance is removed in Shore [2007] where the jump is also defined in every jump ideal containing $\mathbf{0}^{(\omega)}$.

References

Abraham, U. and Shore, R. A. [1986], Initial segments of the Turing degrees of size $\aleph_1$, *Israel J. Math.* **55**, 1-51.

Cai, M. [2012], Array nonrecursiveness and relative recursive enumerability, *J. of Symb. Logic* **77**, 21-32.

Cai, M. and Shore, R. A. [2012], Domination, forcing, array nonrecursiveness and relative recursive enumerability, *J. of Symb. Logic* **77**, 33-48.

Cooper, S. B. [1989], Degrees of unsolvability complementary between recursively enumerable degrees, *Ann. Math. Logic* **4**, 31-73.

Dekker, J. C. E. [1954], A theorem on hypersimple sets, *Proc. Am. Math. Soc.* **5**, 791-796.

Feferman, S. [1965] Some applications of the notions of forcing and generic sets, *Fund. Math.* **56**, 325–345.

Feiner, L. [1970], The strong homogeneity conjecture, *J. Symb. Logic* **35**, 375-377.

Greenberg, N. and Montalbán, A. [2003], Embedding and coding below a 1-generic degree, *Notre Dame J. Formal Logic* **44**, 200-216.

Groszek, M. S. and Slaman, T. A. [1983], Independence results on the global structure of the Turing degrees, *Trans. Am. Math. Soc.* **277**, 579-588.

Hinman, P. G. [1969] Some applications of forcing to hierarchy problems in arithmetic, *Z. Math. Logik Grundlagen Math.* **15**, 341–352.

Jockusch, C. G. jr. [1980], Degrees of generic sets, in *Recursion theory: its generalisation and applications* (Proc. Logic Colloq., Univ. Leeds, Leeds, 1979), *London Math. Soc. Lecture Note Ser.*, **45**, Cambridge Univ. Press, Cambridge-New York, 110–139.

Jockusch, C. G., Jr. and Posner, D. B. [1978], Double jumps of minimal degrees. *J. Symbolic Logic* **43**, 715–724.

Kleene, S. C. and Post, E. L. [1954] The upper semi-lattice of degrees of recursive unsolvability, *Ann. of Math.* (2) **59**, 379–407.

Lerman, M. [1971], Initial segments of the degrees of unsolvability, *Ann. of Math. (2)* **93**, 365-389.

Lerman, M. [1983], *Degrees of Unsolvability*, Springer-Verlag, Berlin.

Martin, D. A. [1966], Classes of recursively enumerable sets and degrees of unsolvability, *Z. Math. Logik Grundlagen Math.* **12**, 295–310.

Miller, W. and Martin, D. A. [1968], The degrees of hyperimmune sets. *Z. Math. Logik Grundlagen Math.* **14**, 159–166.

Nerode, A. and Shore, R. A. [1980], Second order logic and first order theories of reducibility orderings in *The Kleene Symposium*, J. Barwise, H. J. Keisler and K. Kunen, eds., North-Holland, Amsterdam, 181-200.

Nerode, A. and Shore, R. A. [1980a], Reducibility orderings: theories, definability and automorphisms, *Ann. of Math. Logic* **18**, 61-89.

Nies, A., Shore, R. A. and Slaman, T. [1998], Interpretability and definability in the recursively enumerable degrees, *Proceedings of the London Mathematical Society* (3) **77**, 241-291.

Odifreddi, P. [1989] and [1999] *Classical recursion theory*. Vols. I and II. *Studies in Logic and the Foundations of Mathematics* **125** and **143**. North-Holland Publishing Co., Amsterdam.

Odifreddi, P. and Shore, R. A. [1991], Global properties of local degree structures, *Bul. U. Mat. Ital.* **7**, 97-120.

Rogers, H. Jr. [1987], *Theory of recursive functions and effective computability*. Second edition. MIT Press, Cambridge, MA.

Sacks, G. E. [1961], On suborderings of degrees of recursive unsolvability, *Z. Math. Logik Grundlagen Math.* **7**, 46-56.

Sacks, G. E. [1963], Recursive enumerability and the jump operator, *Trans. Am. Math. Soc.* **108**, 223-239.

Shoenfield, J. R. [1959], On degrees of unsolvability, *Ann. Math. (2)* **69**, 644-653.

Shoenfield, J. R. [1960], An uncountable set *of* incomparable degrees, *Proc. Am. Math. Soc.* **11**, 61-62.

Shore, R. A. [1981], The theory of the degrees below $\mathbf{0}'$, *J. London Math. Soc.* (3) **24**, 1-14.

Shore, R. A. [1982], Finitely generated codings and the degrees r.e. in a degree **d**, *Proc. Am. Math. Soc.* **84**, 256-263.

Shore, R. A. [1982a], On homogeneity and definability in the first order theory of the Turing degrees, *J. Symb. Logic* **47**, 8-16.

Shore, R. A. [1988], Defining jump classes in the degrees below $0'$, *Proceedings of the American Mathematical Society* **104**, 287-292.

Shore, R. A. [2007], Direct and local definitions of the Turing jump, *J. of Math. Logic* **7**, 229-262.

Shore, R. A. [2014], Biinterpretability up to double jump in the degrees below $\mathbf{0}'$, *Proc. Am. Math. Soc.* **142**, 351-360.

Shore, R. A. and Slaman, T. A. [1999], Defining the Turing jump, *Math. Research Letters* **6**, 711-722.

Simpson, S. G. [1977], First order theory of the degrees of recursive unsolvability, *Ann. Math. (2)*, **105**, 121-139.

Slaman, T. A. [1991], Degree structures, in *Proc. Int. Cong. Math., Kyoto 1990*, Springer-Verlag, Tokyo, 303-316.

Slaman, T. A.. [2008], Global properties of the Turing degrees and the Turing jump in *Computational prospects of infinity. Part I. Tutorials, Lect. Notes Ser. Inst. Math. Sci. Natl. Univ. Singap.* **14**, World Sci. Publ., Hackensack, NJ, 83–101.

Slaman, T. A. and Steel, J. R. [1989], Complementation in the Turing degrees, *J. of Symb. Logic* **54**, 160-176.

Slaman, T. A. and Woodin, H. W. [1986], Definability in the Turing degrees, *Illinois J. Math.* **30**, 320-334.

Soare, R. I. [1987], *Recursively Enumerable Sets and Degrees*, Springer-Verlag, Berlin.

Spector, C. [1956], On degrees of recursive unsolvability, *Ann. Math. (2)* **64**, 581-592.

Thomason, S. K. [1970], Sublattices and initial segments of the degrees of unsolvability, *Can. J. Math.* **3**, 569-581.

AN INTRODUCTION TO ITERATED ULTRAPOWERS

John Steel*

Department of Mathematics
The University of California, Berkeley
CA 94720-3840 USA
steel@math.berkeley.edu

The basic theory of iterated ultrapowers and their applications is presented. This includes extender-ultrapowers and linear iterations of them. We also develop the pure theory of iteration trees and discuss one of the most important open problems it contains: the Unique Branches Hypothesis (UBH) for countably closed iteration tree on V. Several important applications of iteration trees are also considered.

1. Introduction

In these notes, we develop the basic theory of iterated ultrapowers of models of set theory. The notes are intended for a student who has taken one or two semesters of graduate-level set theory, but may have little or no prior exposure to ultrapowers and iteration.

We shall develop the pure theory, which centers on the question of well-foundedness for the models produced in various iteration processes. In addition, we consider two sorts of applications:

(1) Large cardinal hypotheses yield regularity properties for definable sets of reals. Large cardinal hypotheses yield that logical simple sentences are absolute between V and its generic extensions.
(2) Large cardinal hypotheses admit canonical inner models having wellorders of $\mathbb{R}$ which are simply definable.

---

*The author would like to thank Professor Zhaokuan Hao for his invaluable help in preparing these notes.

Roughly, applications of type (1) involves using the large cardinal hypotheses to construct complicated iterations. Applications of type (2) involves bounding the complexity of the iterations one can produce under a given large cardinal hypothesis.

The notes are organized as follows. In Section 1, we develop the basic theory of ultrapower $\mathrm{Ult}(M, U)$, where M is a transitive model of **ZFC** and U is an ultrafilter over M. In Section 2, we develop the pure theory of iterations of such ultrapowers, and present some applications of type (1). Section 3 concerns applications of type (2). In Section 4 we develop the basic theory of ultrapower $\mathrm{Ult}(M, E)$, where M is as before, but E is now a system of ultrafilters over M known as an "extender". Section 5 concerns linear iteration of such extender-ultrapowers, and its applications of type (1) and (2).

In Section 6, we move from linear iteration to iteration trees, and develop the pure theory of this more general iteration process. This theory is far from complete; indeed, it contains one of the most important question in pure large cardinal theory, the Unique Branches Hypothesis (UBH) for countably closed iteration tree on V. We shall discuss UBH, and its role.

In Section 7, we present some applications of iteration trees in proofs of generic absoluteness and Lebesgue measurability. Those involves Woodin's "extender algebra", and the corresponding "genericity iterations".

Finally, we outline in Section 8 how iteration trees contribute to the theory of canonical inner models with Woodin cardinals.

2. Measures and Embeddings

For this standard material, see for example [[2], Chapter 17].

Let $M \models \mathbf{ZFC}^-$ be transitive, and

$$j : M \to N$$

be elementary. Let $\kappa = \mathrm{crit}(j) =$ least α such that $j(\alpha) \neq \alpha$. We suppose $\kappa \in \mathrm{wfp}(N)$, the well-founded part of N. (Here and later, we assume all well-founded parts have been transitivized.) For $A \subset \kappa$, $A \in M$, put

$$A \in U_j \quad \text{iff} \quad \kappa \in j(A).$$

Then

(1) U_j is a nonprincipal ultrafilter on $\mathcal{P}(\kappa)^M$;

(2) U_j is M κ-complete: if $\langle A_\alpha | \alpha < \beta\rangle \in M$, and $\beta < \kappa$, and each $A_\alpha \in U_j$, then $\bigcap_{\alpha<\beta} A_\alpha \in U_j$;
(3) U_j is M normal: if $f : \kappa \to \kappa$ and $f \in M$, and $f(\alpha) \in \alpha$ for U_j almost every α, then $\exists \beta < \kappa(f(\alpha) = \beta)$ for U_j a.e. α;
(4) $\mathcal{P}(\kappa)^M \subseteq \mathcal{P}(\kappa)^N$. Moreover, $\mathcal{P}(\kappa)^M = \mathcal{P}(\kappa)^N$ iff U_j is M-amenable: whenever $f : \kappa \to \mathcal{P}(\kappa)$ and $f \in M$, then $\{\alpha < \kappa \mid f(\alpha) \in U_j\} \in M$.

Remark 2.1: $M = V$ is an interesting special case. Then $U_j \in M$, so it is certainly amenable. κ is a measurable cardinal in this case.

Exercise 1: Prove (1)-(4) above.

Definition 2.2: An M-**ultrafilter** on κ is an M-amenable nonprincipal ultrafilter on $\mathcal{P}(\kappa)^M$. An M-normal, M-ultrafilter on κ is called an M-**nuf** on κ.

Conversely, suppose $M \models \mathbf{ZFC}^-$ is transitive, and U is a non-principal ultrafilter on $\mathcal{P}(\kappa)^M$. We can, using functions in M, form an **ultrapower**:

$$f \sim g \quad \text{iff} \quad \{\alpha \mid f(\alpha) = g(\alpha)\} \in U_j$$

let $[f] = \{g \mid g \sim f\}$, then

$$[f]\tilde{\in}[g] \quad \text{iff} \quad \{\alpha \mid f(\alpha) \in g(\alpha)\} \in U.$$

We set

$$\mathrm{Ult}(M, U) = (\{[f] \mid f \in M\}, \in),$$

where once again, we assume the well-founded part is transitivized. We then have:

(1) Łoś Theorem: for any $f_0, \ldots, f_n$ and $\varphi(v_0, \ldots, v_n)$,

$$\mathrm{Ult}(M, U) \models \varphi[[f_0], \ldots, [f_n]] \text{ iff for } U \text{ a.e. } \alpha,\ M \models \varphi[f_0(\alpha), \ldots, f_n(\alpha)].$$

(2) $i_U^M : M \to \mathrm{Ult}(M, U)$ is elementary, where

$$i_U^M(x) = [\lambda\alpha.x]$$

the equivalent class of constantly x function.
(3) If U is M-κ-complete, then $\kappa = \mathrm{crit}(i_U^M)$. (In general, the critical point is the M-completeness of U.)
(4) Letting $id : \kappa \to \kappa$ be the identity function, then

$$[f] = i_U^M(f)([id])$$

for any f. (Apply Łoś to see this.) Also for $A \in \mathcal{P}(\kappa)^M$:

$$A \in U \quad \text{iff} \quad [id] \in i_U^M(A).$$

Thus $\mathrm{Ult}(M,U) = \{i_U^M(f)([id]) \mid f \in M\}$ = Skolem closure of $\mathrm{ran}(i_U^M) \cup \{[id]\}$ inside $\mathrm{Ult}(M,U)$.

(5) If U is M-normal, then

$$[id] = \kappa.$$

Thus

$$[f] = i_U^M(f)(\kappa),$$

and

$$A \in U \quad \text{iff} \quad \kappa \in i_U^M(a),$$

and

$$\begin{aligned}\mathrm{Ult}(M,U) &= \{i_U^M(f)(\kappa) \mid f \in M\} \\ &= \text{Skolem closure of } \mathrm{ran}(i_U^M) \cup \{\kappa\} \text{ in } \mathrm{Ult}(M,U).\end{aligned}$$

From (4) above, we easily get

(6) Let U be an M-ultrafilter on κ, and $\langle f_\alpha \mid \alpha < \kappa \rangle \in M$, where $\mathrm{dom}(f_\alpha) = \kappa$ for all α. Then

$$\langle [f_\alpha] \mid \alpha < \kappa \rangle \in \mathrm{Ult}(M,U).$$

Proof: $\langle [f_\alpha] \mid \alpha < \kappa \rangle = \langle i_U^M(f_\alpha)([id]) \mid \alpha < \kappa \rangle$. But

$$\langle i_U^M(f_\alpha) \mid \alpha < \kappa \rangle = i_U^M(\langle f_\alpha \mid \alpha < \kappa \rangle) \restriction \kappa \in \mathrm{Ult}(M,U).$$

So $\mathrm{Ult}(M,U)$ can compute $\langle [f_\alpha] \mid \alpha < \kappa \rangle$ from $[id]$ and $\langle i_U^M(f_\alpha) \mid \alpha < \kappa \rangle$. □

Exercise 2: Prove (1)-(5) above.

Part (5) says that $U = U_{i_U^M}$, that is, U is derived from its own ultrapower embedding. In general, U_j captures only part of the information in j. The relationship is given by the following lemma.

Lemma 2.3: *Let $M \models$ **ZFC**$^-$ be transitive, and $j : M \to N$ be elementary, with $\kappa = \mathrm{crit}(j)$ and $\kappa \in \mathrm{wfp}(N)$. Then we have the commutative diagram*

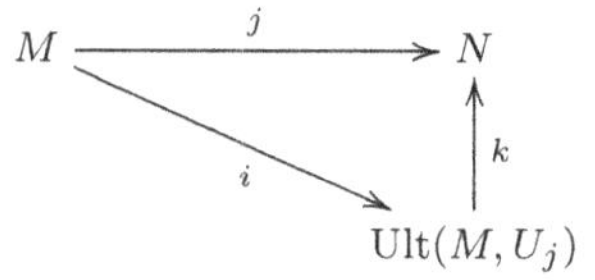

Fig. 1.

where $i = i^M_{U_j}$, *and* $k([f]) = k(i(f)(\kappa)) =_{df} j(f)(\kappa)$. *Moreover,*

(a) $k \restriction \mathcal{P}(\kappa)^{\mathrm{Ult}} = id$ *and* $k \restriction (\kappa^+)^{\mathrm{Ult}} = id$,
(b) if $\mathcal{P}(\kappa)^M = \mathcal{P}(\kappa)^N$, *then* $k \restriction \mathcal{P}(\mathcal{P}(\kappa))^{\mathrm{Ult}} = id$ *and* $k \restriction (\kappa^{++})^{\mathrm{Ult}} = id$.

Exercise 3: Prove Lemma 2.3.

So it follows that $\mathrm{Ult}(M, U_j)$ is isomorphic to $\mathrm{Hull}^N(\mathrm{ran}(j) \cup \{\kappa\})$, via κ.

It is often important to take ultrapowers of models of less than full **ZFC**$^-$. In this case, we only get restricted forms of Łoś' Theorem. For example:

Exercise 4: Let M be transitive, rudimentarily closed, and $M \models$ **AC**. (A paradigm case is $M = L_\lambda$, for some limit λ.) Let U be an ultrafilter on $\mathcal{P}(\kappa)^M$. Then

(a) $\mathrm{Ult}(M, U)$ is well defined,
(b) Łoś' Theorem holds for Σ_0 formulae,
(c) $i^M_U : M \to \mathrm{Ult}(M, U)$ is Σ_1-elementary.

We turn now to the issue of the well-foundedness of the "target model" for our embeddings. We are generally only able to use well-founded models, so this is a crucial issue. We want to stay in the realm of well-founded models!

Note first that if $j : M \to N$ where N is well-founded, then $\mathrm{Ult}(M, U_j)$ is well-founded, since it embeds into N.

A sufficient condition that $\mathrm{Ult}(M, U)$ be well-founded is given by

Definition 2.4: Let U be an ultrafilter on $\mathcal{P}(\kappa)^M$. Then U is **ω_1-complete** iff whenever $A_n \in U$ for all $n < \omega$, then $\bigcap_{n<\omega} A_n \neq \emptyset$.

(There is no requirement that $\langle A_n \mid n < \omega \rangle \in M$, which is why we do not demand $\bigcap_{n<\omega} A_n \in U$.)

Lemma 2.5: *Let* M *be transitive, rudimentarily closed, and* U *be an* ω-*complete ultrafilter on* $\mathcal{P}(\kappa)^M$. *Then* $\mathrm{Ult}(M, U)$ *is well-founded.*

Proof: Suppose $[f_{n+1}]\widetilde{\in}[f_n]$ for all n. Set

$$\alpha \in A_n \iff f_{n+1}(\alpha) \in f_n(\alpha).$$

Then $\bigcap_{n<\omega} A_n = \emptyset$. (Since any element α in this intersection will produce an infinite descending $\in$-sequence in M:

$$f_0(\alpha) \ni f_1(\alpha) \ni f_2(\alpha) \ni \cdots \ni f_n(\alpha) \ni \cdots.)$$ □

Corollary 2.6: *Let M be a transitive, rudimentarily closed, and closed under ω-sequences. Let U be an M-κ-complete ultrafilter on $\mathcal{P}(\kappa)^M$. Then* $\mathrm{Ult}(M,U)$ *is well-founded.*

Proof: If $\langle A_n \mid n<\omega\rangle$ is a counterexample to the well-foundedness, then $\langle A_n \mid n<\omega\rangle \in M$ by the closure of M under ω-sequences. This would then imply U is not M-κ-complete. □

Remark 2.7: Thus $\mathrm{Ult}(V,U)$ is well founded, and more generally, $\mathrm{Ult}(M,U)$ is well-founded when $U \in M$, and $M \models \mathbf{ZFC}$.

Exercise 5: Let U be an ultrafilter on $\mathcal{P}(\kappa)^V$. Then $\mathrm{Ult}(V,U)$ is well-founded iff U is ω-complete.

ω-completeness is only a sufficient condition for the well-foundedness of $\mathrm{Ult}(M,U)$. For if M is itself countable, and U is an ultrafilter on $\mathcal{P}(\kappa)^M$ which is non-principal, then U is not ω-complete. But now take any transitive $N \models \mathbf{ZFC}$ and ultrafilter W on $\mathcal{P}(\kappa)^N$ such that $\mathrm{Ult}(N,W)$ is well-founded. Let

$$\pi : H \to V_\theta$$

where

$$\pi((M,U)) = (N,W)$$

and H is countable transitive. Then π restricts to $\pi : \mathrm{Ult}(M,U) \to \mathrm{Ult}(N,W)$, so $\mathrm{Ult}(M,U)$ is well-founded. But M is countable, so U is not ω-complete.

We conclude this section with a few more basic facts.

Proposition 2.8: *Let U be an M-ultrafilter on κ, then*

$$U \notin \mathrm{Ult}(M,U).$$

Proof: Let $i = i_U^M$, and note that $\mathrm{Ult}(M, U) \models i(\kappa)$ is strongly inaccessible. If $U \in \mathrm{Ult}(M, U)$, then the map

$$f \mapsto [f]_U \qquad (f \in \kappa^\kappa)$$

is in $\mathrm{Ult}(M, U)$, and mps κ^κ onto $i(\kappa)$, contrary to inaccessibility. □

Exercise 6: Let κ be strongly inaccessible. Show any stationary $S \subseteq \kappa$ can be partitioned into κ-many pairwise disjoint stationary sets.

[Hint: Otherwise, we get a stationary $T \subseteq S$ such that if U is the club filter on T, then U is a V-ultrafilter on κ. Now show $U \in \mathrm{Ult}(V, U)$.]

Proposition 2.9: *Let U be a V-nuf on κ, and $i = i_M^V$. Then*

(a) $2^\kappa < i(\kappa) < (2^\kappa)^+$*;*
(b) if $\mathrm{cof}(\alpha) \neq \kappa$*, then* $i(\alpha) = \sup_{\beta<\alpha} i(\beta)$*;*
(c) if $\mathrm{cof}(\alpha) \neq \kappa$ *and* $\forall\beta < \alpha(\beta^\kappa < \alpha)$*, then* $i(\alpha) = \alpha$*.*

Proof: For (a), note $2^\kappa \leq (2^\kappa)^{\mathrm{Ult}(V,U)} < i(\kappa)$, because $\mathcal{P}(\kappa) \subseteq \mathrm{Ult}(V, U)$, and $i(\kappa)$ is inaccessible there, But $f \to [f]_U$ $(f \in \kappa^\kappa)$ shows $|i(\kappa)| \leq 2^\kappa$.

For (b), if $[f] < i(\alpha)$, then $f(\nu) < \alpha$ for U a.e. ν. So $[f] < i(\beta)$.

(c) is also easy. □

Exercise 7: Prove Rowbottom's Theorem: Let U be a M-nuf on κ, where $M \models \mathbf{ZFC}$ is transitive. (M rudimentarily closed sufficient.) Let $f : [f]^n \to \gamma$, with $\gamma < \kappa$, and $f \in M$. Then there is an $A \in U$ such that f is constant on $[A]^n$.

[Hint: The proof is by induction on n. You need to prove a little more than what is stated in order to cope with the possibility that $U \notin M$.]

3. Iterated Ultrapowers

Let $M \vdash \mathbf{ZFC}^-$ be transitive, and suppose

$$M \models \mathcal{E} \text{ is a set of normal ultrafilters.}$$

For any $U \in \mathcal{E}$, we can form $M_1 = \mathrm{Ult}(M, U)$, with $i_U^M : M \to M_1$ elementary. We can then take any $W \in i_E^M(\mathcal{E})$ and form $\mathrm{Ult}(M_1, W) = M_2$, with $i_W^{M_1} : M_1 \to M_2$ elementary. (That is, we can do so if M_1 was well-founded. It would make perfect sense for ill-founded M_1, as well, but formally speaking, we have not consider that case.) In this way, we produce

$$M = M_0 \to M_1 \to M_2 \to \cdots \to M_\alpha \to M_{\alpha+1} \to \cdots$$

where we continue at limit steps by taking M_λ to be the direct limit of the M_α for $\alpha < \lambda$.

Definition 3.1: Let M be transitive, and

$$M \models \mathbf{ZFC}^- + \text{“}\mathcal{E} \text{ is a set of nufs”}.$$

A **linear iteration** of $(M, \mathcal{E})$ is a sequence $I = \langle U_\alpha \mid \alpha < \beta \rangle$ such that there are (unique) transitive M_α, $\alpha < \beta$, and $i_{\alpha\gamma} : M_\alpha \to M_\gamma$ for $\alpha \leq \gamma < \beta$, with

(1) $M_0 = M$;
(2) $U_\alpha \in i_{0\alpha}(\mathcal{E})$ for $\alpha < \beta$;
(3) if $\alpha + 1 < \beta$, then

$$\begin{aligned} M_{\alpha+1} &= \text{Ult}(M_\alpha, U_\alpha), \\ i_{\alpha,\alpha+1} &= i_{U_\alpha}^{M_\alpha}, \text{ and} \\ i_{\xi,\alpha+1} &= i_{\alpha,\alpha+1} \circ i_{\xi,\alpha} \text{ for } \xi < \alpha. \end{aligned}$$

(4) if $\lambda < \beta$ is a limit,

$$\begin{aligned} M_\lambda &= \text{direct limit of } M_\alpha,\ \alpha < \lambda, \text{ under } i_{\alpha\gamma}\text{'s} \\ i_{\alpha\lambda} &= \text{direct limit map, for } \alpha < \lambda. \end{aligned}$$

If I is a linear iteration of $(M, \mathcal{E})$, we write U_α^I, M_α^I, $i_{\alpha\gamma}^I$ for the associate ultrafilters , models, and embeddings. There is a unique “last model” associated to I:

$$\begin{aligned} M_\infty^I &= \text{direct limit of } M_\alpha^I, \text{ for } \alpha < \text{lh}(I), \text{ if lh}(I) \text{ is a limit.} \\ M_\infty^I &= \text{Ult}(M_\alpha^I, U_\alpha^I),\text{s if } \alpha + 1 = \text{lh}(I). \end{aligned}$$

We let $I_{\alpha,\infty}^I : M_\alpha^I \to M_\infty^I$ be the canonical embedding, for $\alpha < \text{lh}(I)$. Unlike the M_α^I, M_∞^I may be ill-founded.

Definition 3.2: Let $M \models \mathbf{ZFC}^- + \text{“}\mathcal{E}$ is a set of nufs”, M transitive. We say $(M, \mathcal{E})$ is **linearly iterable** iff for every linear iteration I of $(M, \mathcal{E})$, M_∞^I is well-founded.

Just as it sufficed for well-foundedness of single ultrapower, ω-completeness suffices for linear iterability.

Theorem 3.3: *Let $M \models \mathbf{ZFC}^- + \text{“}\mathcal{E}$ is a set of nufs”, with M transitive. Suppose every $U \in \mathcal{E}$ is ω-complete (in V). Then $(M, \mathcal{E})$ is linearly iterable.*

Remark 3.4: Suppose U is a nuf on κ, and set $M_0 = V$, and $M_{n+1} = \mathrm{Ult}(M_n, U)$ for all n. (By induction, $\mathcal{P}(\kappa)^{M_n} = \mathcal{P}(\kappa)$, so this make sense.) Then the direct limit of M_n is ill-founded, as the picture shows:

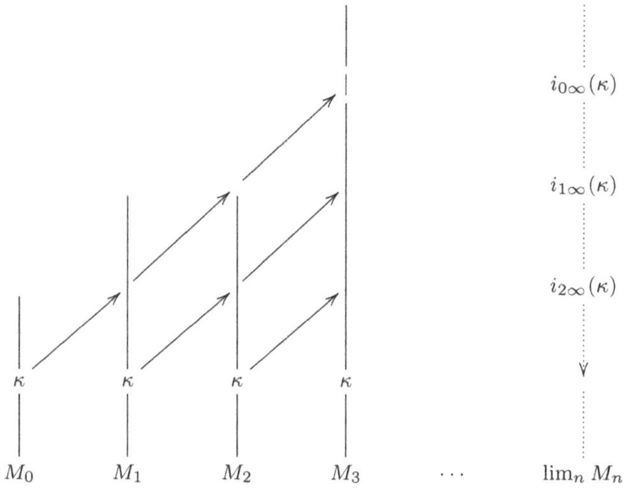

Fig. 2.

The moral is that in a legitimate iteration, you cannot just pull the next ultrafilter out of your hat! It has to come from the last model.

The proof of Theorem 3.3 will rely on some lemmas of independent interest. Let us call $(M, \mathcal{E})$ s.t. $M \models \mathbf{ZFC}^- +$ "$\mathcal{E}$ is a set of nufs", and M transitive, a **good pair**. We say $(M, \mathcal{E})$ is α-**linearly iterable** iff whenever I is a linear iteration of $(M, \mathcal{E})$ and $\mathrm{lh}(I) < \alpha$, then M^I_∞ is well-founded.

Lemma 3.5: *Let $(M, \mathcal{E})$ be a good pair which is α-linearly iterable. Let $\pi : N \to M$ elementary, with $\pi(\mathcal{F}) = \mathcal{E}$. Then $(N, \mathcal{F})$ is α-linearly iterable.*

Proof: Let I be an iteration of $(N, \mathcal{F})$ with $\mathrm{lh}(I) < \alpha$. We can complete the diagram

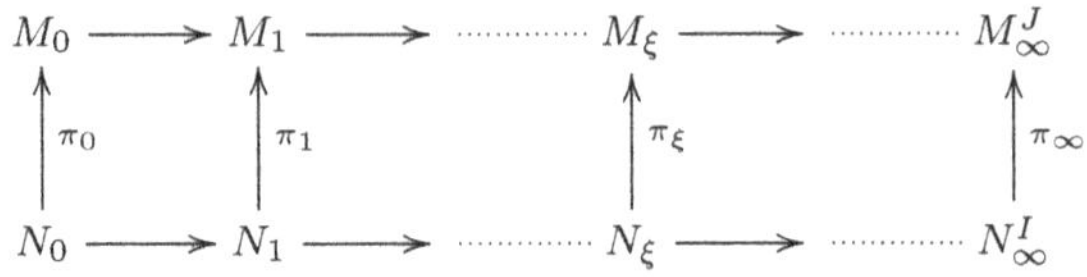

Fig. 3.

Here $M_0 = M$, $N = N_0$, $\pi_0 = \pi$, and $U_\xi = \pi_\xi(U^I_\xi)$ give us J. We get $\pi_{\xi+1}$ by setting

$$\pi_{\xi+1}([f]_{U^I_\xi}) = [\pi_\xi(f)]_{U^J_\xi},$$

and π_λ a limit because the diagram commutes. Since M_∞ is well founded, and N^I_∞ embeds into it, N^I_∞ is well-founded. □

Failures of iterability reflect into the countable:

Lemma 3.6: *Let $(M, \mathcal{E})$ be a good pair. Equivalent are:*

(1) $(M, \mathcal{E})$ is linearly iterable;
(2) whenever $\pi : N \to M$ with N countable and $\pi(\mathcal{F}) = \mathcal{E}$, then $(N, \mathcal{F})$ is ω_1-linearly iterable.

Proof: Lemma 3.5 gives (1)→(2). Now assume (1) fails, and let I be an iteration of $(M, \mathcal{E})$ such that M^I_∞ is ill-founded. Let $\sigma : H \to V_\theta$ with H countable transitive, θ large, and $\sigma((N, \mathcal{F})) = (M, \mathcal{E})$, and $\sigma(J) = I$. Then

$$H \models J \text{ is an iteration of } (N, \mathcal{F}) \text{ with } M^J_\infty \text{ ill-founded.}$$

But the right hand side is absolute for well-founded models, so as $\mathrm{lh}(J) < \omega_1$, $(N, \mathcal{F})$ is not ω_1-linearly iterable. But setting $\pi = \sigma \restriction N$, this shows (2) fails. □

The following lemma is the beginning of our proof of Theorem 3.3. It is due to Jensen.

Lemma 3.7: *Let $M \models$ **ZFC**$^-$ + "U is a nuf on κ", with M transitive. Equivalent are:*

(1) U is ω-complete;
(2) whenever $\pi : N \to M$ is elementary, with N countable and $\pi(W) = U$, then there is a σ such that

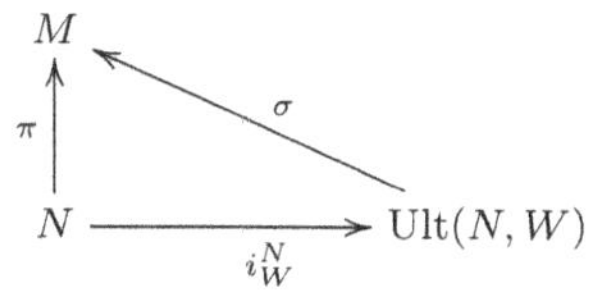

Fig. 4.

commutes.

Proof: (1)→ (2). Let $\pi : N \to M$ with $\pi(W) = U$. Pick a "typical object" for ran(π), that is α such that

$$\alpha \in \bigcap_{A \in W} \pi(A).$$

This we can do because U is ω-complete. Now set

$$\sigma([f]_M^N) = \pi(f)(\alpha).$$

□

Exercise 8: Show σ is well defined, elementary, and $\pi = \sigma \circ i_W^N$.

Exercise 9: Prove (2) → (1).

The map σ in (2) of Lemma 3.7 is called "π-realization" of Ult(N, W). The Lemma 3.7 says that if U is ω-complete, then countable fragment of Ult(M, U) can be "realized" back in M.

Proof: [Proof of Theorem 3.3] Let $(M, \mathcal{E})$ be a good pair, and every $U \in \mathcal{E}$ be ω-complete. Suppose $(M, \mathcal{E})$ is not linearly iterable. Let, by Lemma 3.6, $\pi : N \to M$ with N countable, and $\pi(\mathcal{F}) = \mathcal{E}$, and I a countable iteration of $(N, \mathcal{F})$ such that M_∞^I is ill-founded. Repeatedly using Lemma 3.7, we get

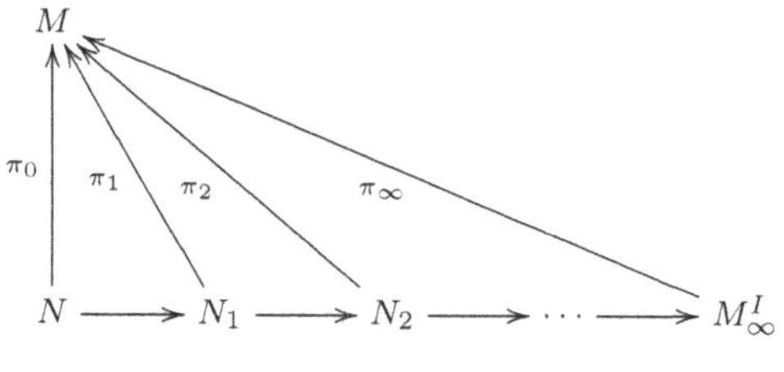

Fig. 5.

(π_λ for λ limit comes from the commutativity of the diagram.) Since M is well-founded, so is M^I_∞, a contradiction. □

Corollary 3.8: (ZFC) *If $\mathcal{E}$ is a set of nufs, then $(V, \mathcal{E})$ is linearly iterable.*

Corollary 3.9: *Let $M \models$ **ZFC**+ "$\mathcal{E}$ is a family of nufs", with M transitive, and $\omega_1 \in M$. Then $(M, \mathcal{E})$ is linearly iterable.*

Proof: [Proof Sketch] Working inside M, where Corollary 3.8 holds, construct a "universal" linear iteration $I = \langle U_\alpha \mid \alpha < \lambda \rangle$ with (a) $\mathrm{cof}(\lambda) = \omega$, and (b) whenever $W \in i_{0\alpha}(\mathcal{E})$, then $i_{\alpha\beta}(W) = U_\beta$ for cofinally many β. This implies that every $W \in i_{0\infty}(\mathcal{E})$ is ω-complete. Thus $(M^I_\infty, i_{0\infty}(\mathcal{E}))$ is linearly iterable in V. Since $i_{0\infty} : M \to M^I_\infty$, $(M, \mathcal{E})$ is linearly iterable in V. □

Exercise 10: Prove that every $W \in i_\infty(\mathcal{E})$ is ω-complete, granted (a) and (b).

Exercise 11: Where is $M \models$ **ZFC** (rather than just $M \models$ **ZFC**$^-$) used in the proof of Corollary 3.9?

Some Applications

A. Regularity properties of definable sets of reals

Theorem 3.10: (Gaifman, Rowbottom) *If there is a measurable cardinal, then for all reals x, $\omega_1^{L[x]}$ is countable.*

Proof: Let U be a nuf on κ. Clearly, there is an M such that $(M, \{U\})$ is a good pair. By Theorem 3.3, $(M, \{U\})$ is iterable. Now let $x \in \mathbb{R}$, and let $\pi : N \to M$ be elementary, with N countable transitive, $x \in N$, and $\pi(W) = U$. So $(N, \{W\})$ is a good pair, and iterable. Let

$$\alpha = (\omega_1^{L[x]})^N.$$

It is enough to see $\alpha = \omega_1^{L[x]}$. But let I be the unique linear iteration of N of length ω_1, and $i : N \to N^I_\infty$ the canonical embedding. Then

$$\alpha = i(\alpha) = (\omega_1^L[x])^{N^I_\infty}.$$

Since $\omega_1 \subseteq N^I_\infty$, this implies $\alpha = \omega_1^{L[x]}$. □

Theorem 3.11: (Solovay) *If there is a measurable cardinal, then all $\utilde{\Sigma}^1_2$ sets of reals are Lebesgue measurable, have the Baire Property, and have the Perfect Set Property.*

Proof: See [2]. The proof uses Theorem 3.10. □

Determinacy is the fundamental regularity property, and we have

Theorem 3.12: (Martin) *If there is a measurable cardinal, then all* $\underset{\sim}{\Pi}^1_1$ *games are determined.*

A proof using iterated ultrapowers can be given, but we shall not go that far afield now.

B. Correctness and generic absoluteness

Theorem 3.13: *Let* $(M, \{U\})$ *be a good pair which is linearly iterable. Then*

$$(HC^M, \in) \prec_{\Sigma_1} (HC, \in).$$

Proof: Here $HC = \{x \mid |TC(x)| < \omega_1\}$ is the class of hereditarily countable sets. Σ_1-over-HC is equivalent to Σ^1_3.

Let I be the unique iteration of M of length ω_1. Then

$$HC^M = HC^{M^I_\infty} \prec_{\Sigma_1} HC,$$

using $\omega_1 \subset M^I_\infty$ and Shoenfield absoluteness. □

Theorem 3.14: (Martin, Solovay) *Let* κ *be measurable, and* G *be* $\mathbb{P}$*-generic for some* $\mathbb{P}$ *with* $|\mathbb{P}| < \kappa$*. Then*

$$(HC, \in)^V \prec_{\Sigma_2} (HC, \in)^{V[G]}.$$

Proof: By Tarski-Vaught, it is enough to see that if $x \in HC^V$, and

$$p \Vdash^{\mathbb{P}} (HC, \in) \models \varphi[\check{x}, \tau],$$

where φ is Π_1, then for some $y \in V$, $(HC, \in)^V \models \varphi[x, y]$. But by Löwenheim-Skolem, we can get an iterable good pair $(N, \{U\})$ such that N is countable, and (for q, $\mathbb{Q}$, σ the collapses of p, $\mathbb{P}$, τ)

$$N \models (q \Vdash^{\mathbb{Q}} (HC, \in) \models \varphi[\check{x}, \sigma]).$$

Let $i : N \to N^I_\infty$ be an iteration map, with $\omega_1 \subseteq N^I_\infty$. We may assume $N \models$ "V_x exists", and $\mathbb{Q} \in V^N_\alpha$, where $\alpha < \kappa$. It follows that $V^N_\alpha = V^{N^I_\infty}_\alpha$ is countable, and hence there is in V a $\mathbb{Q}$-generic g over N^I_∞ such that $q \in g$. Note here that $i(\langle q, \mathbb{Q}, \sigma\rangle) = \langle q, \mathbb{Q}, \sigma\rangle$. Thus $(HC, \in)^{N^I_\infty[g]} \models \varphi[\check{x}, \sigma_g]$. Since

φ is Π_1 and $\omega_1 \subseteq N^I_\infty$, Shoenfield absoluteness implies $(HC, \in)^V \models \varphi[\check{x}, \sigma_g]$. So σ_g is the desired y. □

We conclude this section with exercises on two basic features of linear iteration.

What is the analog of $\text{Ult}(M, U) = \{i^M_U(f)(\kappa) \mid f \in M\}$, for $\kappa = \text{crit}(U)$? We can generate M^I_∞ from $\text{ran}(i^I_{0,\infty})$ together with all the critical points, as follows:

Exercise 12: Let $I = \langle U_\alpha \mid \alpha < \beta \rangle$ be a linear iteration of M, with $\kappa_\alpha = \text{crit}(U_\alpha)$, and set $C = \{i_{\alpha+1,\infty}(\kappa_\alpha) \mid \alpha < \beta\}$. (Note the "$\alpha+1$" here!) Then

$$M^I_\infty = \{i_{0,\infty}(f)(a) \mid a \in C^{<\omega}\}.$$

Picture:

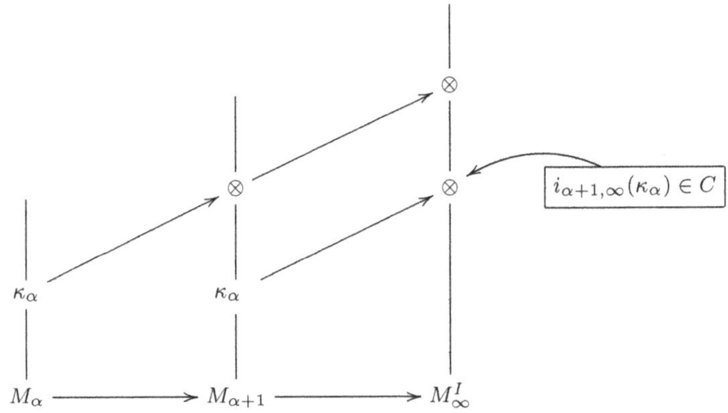

Fig. 6.

Notice that every $\gamma \in C$ comes from a unique stage in I, in fact, $\gamma = i_{\alpha+1,\infty}(\kappa_\alpha)$ for α least such that $\gamma \in \text{ran}(i_{\alpha+1,\infty})$. (Simply because $\kappa_\alpha \notin \text{ran}(I_{\alpha,\alpha+1})$.) Ordinals in C "belonging to the same measure" are indiscernible, in the following sense:

Exercise 13: Let $I = \langle U_\alpha \mid \alpha < \beta \rangle$ be a linear iteration of M, and $\kappa_\alpha = \text{crit}(U_\alpha)$. For $U \in M$, put

$$C_u = \{i_{\alpha+1,\infty}(\kappa_\alpha) \mid U_\alpha = i_{0\alpha}(U)\}.$$

Let $\gamma_0 < \cdots < \gamma_{n-1}$ and $\delta_0 < \cdots < \delta_{n-1}$ with each $\gamma_i, \delta_i \in C_U$. Let $t \in \mathrm{ran}(i^I_{0,\infty})$. Then

$$M^I_\infty \models \varphi[\gamma_0, \ldots, \gamma_{n-1}, t] \iff M^I_\infty \models \varphi[\delta_0, \ldots, \delta_{n-1}, t]$$

for all wff $\varphi(v_0, \ldots, v_{n-1})$.

[Hint: Let $\gamma_i = i_{\alpha_i+1,\infty}(\kappa_{\alpha_i})$. Let J be the iteration of M of length n where you just hit U and its images. Consider the diagram

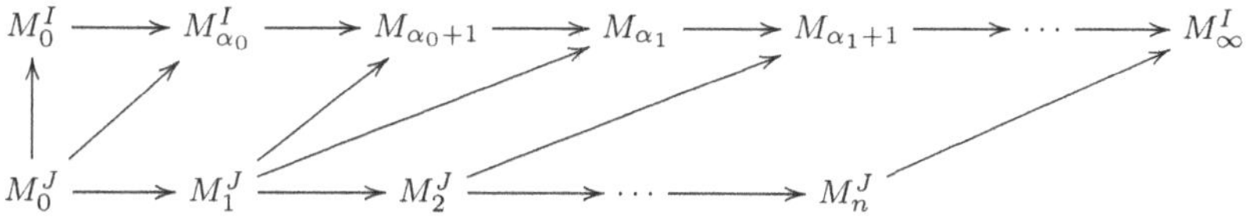

Fig. 7.

Make sense of the diagram, and use it to show that if $v_i = \mathrm{crit}(U^J_i)$ for $i < n$, then $M^I_\infty \models \varphi[\gamma_0, \ldots, \gamma_{n-1}, i_{0\infty}(U)]$ iff $M^J_n \models \varphi[\nu_0, \ldots, \nu_{n-1}, i_{0n}(U)]$. Since a similar equivalence for the δ_i's, we are done.]

4. Canonical Inner Models and Comparison

Some of the main applications of iterated ultrapowers lie in inner model theory.

We begin with models constructed from coherent sequences of normal ultrafilters. ([5]) Roughly speaking, a "coherent sequence" is one in which the nufs occur in order of strength, without leaving gaps. The strength order is the "Mitchell order" $\lhd$, where for U and W nufs,

$$U \lhd W \quad \text{iff} \quad U \in \mathrm{Ult}(V, W).$$

Clearly, $U \lhd W \to \mathrm{crit}(U) \leq \mathrm{crit}(W)$, and $\mathrm{crit}(U) < \mathrm{crit}(W) \to U \lhd W$. Thus the interesting case is when $\mathrm{crit}(U) = \mathrm{crit}(W)$. By Proposition 1.6, $U \ntriangleleft U$. In fact,

Lemma 4.1: $\lhd$ *is well-founded.*

Proof: Let $U \lhd W$ and $\mathrm{crit}(U) = \mathrm{crit}(W) = \kappa$. Then

$$i^V_U(\kappa) = i^{\mathrm{Ult}(V,W)}_U(\kappa) < i^V_W(\kappa).$$

The first equality holds because V and $\mathrm{Ult}(V, W)$ have the same $f : \kappa \to \kappa$. The inequality holds because $i^V_W(\kappa)$ is inaccessible in $\mathrm{Ult}(V, W)$.

So the map $U \mapsto i^V_U(\kappa)$ maps $\{U \mid \mathrm{crit}(U) = \kappa\}$ into the ordinals, witness $\lhd$ is well-founded. □

The argument actually shows $\lhd \restriction \{U \mid \mathrm{crit}(U) = \kappa\}$ has rank $< (2^\kappa)^\kappa$.

It is consistent that $\lhd$ is not linear, in fact, there can be $2^{(2^\kappa)}$ nufs on κ which are $\lhd$-minimal. ([3]) But in the canonical inner models, $\lhd$ is linear.

Definition 4.2: A **coherent sequence of nufs** is a function $\mathcal{U}$ such that $\mathrm{dom}(\mathcal{U}) \subseteq \mathrm{Ord} \times \mathrm{Ord}$, and

(1) $(\kappa, \beta) \in \mathrm{dom}(\mathcal{U}) \Longrightarrow \mathcal{U}(\kappa, \beta)$ is a nuf on κ;
(2) $(\kappa, \beta) \in \mathrm{dom}(\mathcal{U}) \wedge \gamma < \beta \Longrightarrow (\kappa, \gamma) \in \mathrm{dom}(\mathcal{U})$,
(3) Letting $o^{\mathcal{U}}(\kappa) = \sup\{\beta \mid (\kappa, \beta) \in \mathrm{dom}(\mathcal{U})\}$, $i : V \to \mathrm{Ult}(V, \mathcal{U}(\kappa, \beta))$, we have $o^{i(\mathcal{U})}(\kappa) = \beta$, and

$$i(\mathcal{U})(\kappa, \gamma) = U(\kappa, \gamma), \quad \text{for all } \gamma < \beta.$$

Remark 4.3: Let's write $\mathcal{U} \restriction (\kappa, \gamma)$ for

$$\mathcal{U} \restriction \{(\alpha, \tau) \mid \alpha < \kappa \text{ or } (\alpha = \kappa \wedge \tau < \gamma)\}.$$

Condition (3) can then be expressed

$$i^V_{\mathcal{U}(\kappa,\beta)}(\mathcal{U}) \restriction (\kappa, \beta + 1) = \mathcal{U} \restriction (\kappa, \beta).$$

Picture:

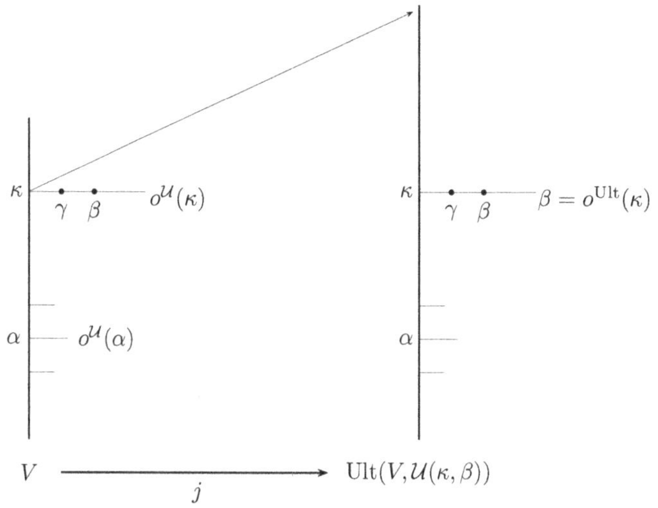

Fig. 8.

$$i(\mathcal{U})(\alpha,\gamma) = \begin{cases} \mathcal{U}(\alpha,\gamma), & \text{if } \alpha < \kappa, \text{ or } \alpha = \kappa \wedge \gamma < \beta; \\ \text{undefined}, & \text{if } \alpha = \kappa \wedge \gamma \geq \beta. \end{cases}$$

For $\mathcal{U}$ a coherent segment of nufs, we set

$$L[\mathcal{U}] = L[A],$$

where

$$A = \{(\beta,\gamma,X) \mid X \in \mathcal{U}(\beta,\gamma)\}\,.$$

This gives us that $L[\mathcal{U}] = \mathbf{ZFC} +$ "$V = L[\mathcal{U}]$", and $\mathcal{U}(\alpha,\beta) \cap L[\mathcal{U}] \in L[\mathcal{U}]$ for all (β,γ) in dom($\mathcal{U}$). It follows that

$$L[\mathcal{U}] \models \mathcal{U}(\beta,\gamma) \text{ is a nuf on } \beta$$

for all $(\beta,\gamma) \in \mathrm{dom}(\mathcal{U})$, where on the right we write "$\mathcal{U}(\beta,\gamma)$" for "$\mathcal{U}(\beta,\gamma) \cap L[\mathcal{U}]$", as we shall do when context makes the meaning clear.

It is still open (almost 40 years after [5]) whether

$$L[\mathcal{U}] \models \mathcal{U} \text{ is a coherent sequence of nufs.}$$

(Later developments reduced the importance of this question.) The problem is that ultrapowers computed in $L[\mathcal{U}]$ may diverge too much from those computed in V. However, [5] did show

Theorem 4.4: *Suppose there is an elementary* $j : V \to M$ *with* $\mathrm{crit}(j) = \kappa$ *and* $V_{\kappa+2} \subseteq M$. *Then there is a coherent segment* U *of nufs such that*

$$L[\mathcal{U}] \models U \text{ is coherent } \wedge \exists\alpha(o^U(\alpha) = \alpha^{++}).$$

Proof: [Proof Sketch] Let κ, j and M be as in the hypothesis. We define a "maximal coherent sequence below $o(\alpha) = \alpha^{++}$ as follows.

Suppose we have defined $\mathcal{U} \restriction (\alpha, \beta)$, where $\alpha < \kappa$, so that $L[\mathcal{U} \restriction (\alpha, \beta)] \models \mathcal{U} \restriction (\alpha, \beta)$ is coherent. If $o(\alpha) = \alpha^{++}$ holds in $L[\mathcal{U} \restriction (\alpha, \beta)]$, we have the desired $\mathcal{U}$, and we stop the construction. So suppose $o(\alpha) < \alpha^{++}$ in $L[\mathcal{U} \restriction (\alpha, \beta)]$. Now pick a nuf W on α such that

$$L[\mathcal{U} \restriction (\alpha, \beta)^\frown \langle W \rangle] \models \mathcal{U} \restriction (\alpha, \beta)^\frown \langle W \rangle \text{ is coherent}$$

and set

$$\mathcal{U}(\alpha, \beta) = W,$$

if there is such a W. (**AC** is used here.) If there is no such W, we set $o^{\mathcal{U}}(\alpha) = \beta$, and go on to defining $\mathcal{U}(\alpha', 0)$ for some $\alpha' > \alpha$.

This defines our $\mathcal{U}$, with $\mathrm{dom}(\mathcal{U}) \subseteq \kappa \times \kappa$. It is enough to see

Claim. The construction reaches some $(\alpha, \beta) \in \kappa \times \kappa$ such that $L[\mathcal{U} \restriction (\alpha, \beta)] \models o^{\mathcal{U}}(\alpha) = \beta = \alpha^{++}$.

Proof: Suppose not. Now work in M, where $j(\mathcal{U})$ is a maximal sequence below $j(\kappa)$. Let $\beta = o^{j(\mathcal{U})}(\kappa)$, and let

$$W = U_j = \{A \subseteq \kappa \mid \kappa \in j(A)\}.$$

It is enough to show

$$L[j(\mathcal{U}) \restriction (\kappa, \beta)^\frown \langle W \rangle] \models j(\mathcal{U}) \restriction (\kappa, \beta)^\frown \langle W \rangle \text{ is coherent,} \qquad (*)$$

for then $j(\mathcal{U})$ is not maximal in M. Note here the crucial fact: $W \in V_{\kappa+2}$, so $W \in M$! This is where we use the strength of j.

To prove $(*)$, we consider the diagram

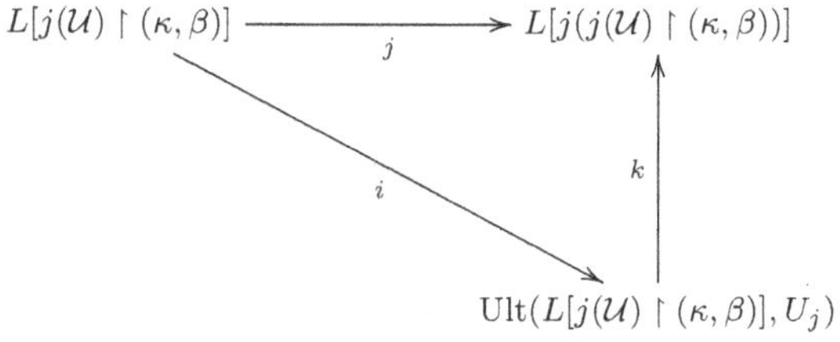

Fig. 9.

□

Exercise 14:

(a) Give precise definitions of i and k;
(b) Show $j(\mathcal{U}) \restriction (j(\kappa), 0) = j(j(\mathcal{U})) \restriction (j(\kappa), 0)$;
(c) Show $k \restriction (\beta + 1) =$ identity;
(d) Prove $(*)$.

This complete our sketch of the proof of Theorem 4.4.
Thus, granted large cardinals in V, there are inner models $L[\mathcal{U}]$ such that $L[\mathcal{U}] \models \mathcal{U}$ is coherent, and $L[\mathcal{U}] \models$ "There are many measurable cardinals". We now show such models are canonical, for example, every real number in such a model is ordinal definable in simple way.

Definition 4.5: A **measures-premouse** is a pair $(M, \mathcal{U})$ such that

$$M \models \mathbf{ZFC}^- + \text{“}\mathcal{U} \text{ is a coherent sequence of nufs”} + \text{“}V = L[\mathcal{U}]\text{”}.$$

A **measures-mouse** is a linearly iterable measures premouse.

Notice that a measures-premouse is a good pair, after we forget the order on $\mathcal{U}$. So linear iterability for it makes sense, and our earlier results apply. If I is an iteration of $\mathfrak{M} = (M, \mathcal{U})$, then we write $\mathfrak{M}^I_\alpha = (M^I_\alpha, i^I_{0\alpha}(\mathcal{U}))$ and $\mathfrak{M}^I_\infty = (M^I_\infty, i^I_{0\infty}(\mathcal{U}))$.

Definition 4.6: Let $\mathfrak{M} = (M, \mathcal{U})$ and $\mathfrak{N} = (N, \mathcal{W})$ be measures-premouse. We say that $\mathfrak{M}$ is an **initial segment of** $\mathfrak{N}$, and write $\mathfrak{M} \trianglelefteq \mathfrak{N}$, iff for all $\kappa \in M$,

(a) $o^{\mathcal{U}}(\kappa) = o^{\mathcal{W}}(\kappa)$, and
(b) $\mathcal{U}(\kappa, \beta) = \mathcal{W}(\kappa, \beta) \cap M$ for all $\beta < o^{\mathcal{U}}(\kappa)$,
(c) $\mathrm{Ord}^{\mathfrak{M}} \leq \mathrm{Ord}^{\mathfrak{N}}$.

Remark 4.7: Let $A_{\mathcal{U}} = \{(\beta, \gamma, X) \mid X \in \mathcal{U}(\beta, \gamma)\}$, and similiarly for $A_{\mathcal{W}}$. So $M = L_\alpha[A_{\mathcal{U}}]$ for some α. Clauses (a)-(c) say that $\alpha \leq \mathrm{Ord}^{\mathfrak{N}}$, and

$$(L_\alpha[A_{\mathcal{U}}], \in, A_{\mathcal{U}}) = (L_\alpha[A_{\mathcal{W}}], \in, A_{\mathcal{W}} \cap L_\alpha[A_{\mathcal{W}}]).$$

The key to inner model theory at any level is a **comparison process**, a method by which two mice can be simultaneously iterated so that an iteration of one is an initial segment of an iteration of the other. At the level of measures-mice, linear iteration suffices for comparison, and we get

Lemma 4.8: (Comparison Lemma for Measures-Mice) *Let $\mathfrak{M}$ and $\mathfrak{N}$ be measures-mice which are sets. Then there are linear iteration I and J such that $\mathfrak{M}^I_\infty \trianglelefteq \mathfrak{N}^J_\infty$ or $\mathfrak{N}^J_\infty \trianglelefteq \mathfrak{M}^I_\infty$.*

Remark 4.9: There is a version of Lemma 4.8 which holds for proper class $\mathfrak{M}$, $\mathfrak{N}$. Then I and J might be a proper class.

Proof: We define initial segment I_ν and J_ν of I and J, by induction on ν. Set $I_0 = \emptyset = J_0$. Let

$$\mathfrak{M}^{I_\nu}_\infty = (P, \mathcal{H})$$

and

$$\mathfrak{N}^{J_\nu}_\infty = (Q, \mathcal{L})$$

be the two last models, where at stage $\nu = 0$ we set $\mathfrak{M}^{I_0} = \mathfrak{M}$ and $\mathfrak{N}^{I_0} = \mathfrak{N}$. We may assume $(P, \mathcal{H}) \ntrianglelefteq (Q, \mathcal{L})$ and $(Q, \mathcal{L}) \ntrianglelefteq (P, \mathcal{H})$, and otherwise we can set $I = I_\nu$ and $J = J_\nu$, and our comparison has succeeded.

We now obtain $I_{\nu+1}$ and $J_{\nu+1}$ by **iterating away the least disagreement** between $(P, \mathcal{H})$ and $(Q, \mathcal{L})$. Namely, let (κ, β) be lexicographically least such that either

(a) $(\kappa, \beta) \in \mathrm{dom}(\mathcal{H}) \bigtriangleup \mathrm{dom}(\mathcal{L})$,

or

(b) $(\kappa, \beta) \in \mathrm{dom}(\mathcal{H}) \cap \mathrm{dom}(\mathcal{L})$, and $\mathcal{H}(\kappa, \beta) \cap P \cap Q \neq \mathcal{L}(\kappa, \beta) \cap P \cap Q$.

If (b) holds, then we set

$$I_{\nu+1} = I_\nu {}^\frown \langle \mathcal{H}(\kappa, \beta) \rangle,$$
$$J_{\nu+1} = J_\nu {}^\frown \langle \mathcal{L}(\kappa, \beta) \rangle.$$

If (a) holds, and $(\kappa, \beta) \in \text{dom}(\mathcal{H})$, we set

$$I_{\nu+1} = I_\nu \frown \langle \mathcal{H}(\kappa, \beta) \rangle,$$
$$J_{\nu+1} = J_\nu.$$

If (a) holds, and $(\kappa, \beta) \in \text{dom}(\mathcal{L})$, we set

$$I_{\nu+1} = I_\nu,$$
$$J_{\nu+1} = J_\nu \frown \langle \mathcal{L}(\kappa, \beta) \rangle.$$

This defines $I_{\nu+1}$ and $J_{\nu+1}$. For λ a limit, we let $I_\lambda = \bigcup_{\nu<\lambda}$ and $J_\lambda = \bigcup_{\nu<\lambda} J_\nu$.

It is enough to show our process terminate. In fact

Claim 1: For some $\nu < \max(|\text{TC}(M)|, |\text{TC}(N)|)^+$, $\mathfrak{M}_\infty^{I_\nu} \trianglelefteq \mathfrak{N}_\infty^{J_\nu}$ or $\mathfrak{N}_\infty^{J_\nu} \trianglelefteq \mathfrak{M}_\infty^{I_\nu}$.

Proof: [Proof of the Claim] Let $\theta = \max(|\text{TC}(M)|, |\text{TC}(N)|)^+$. It is easy to see $\theta = \text{dom}(I_\theta) = \text{dom}(J_\theta)$. Let us write M_α, $i_{\alpha\beta}$ for the models and embeddings of I_θ, and N_α and $j_{\alpha\beta}$ for those of J_θ. Now let

$$\pi : S \to V_\gamma,$$

where $\gamma >> \theta$, S is transitive, $|S| < \theta$, everything relevant is in $\text{ran}(\pi)$. We can arrange that for some $\alpha < \theta$,

$$\pi(\alpha) = \theta,$$

and

$$\pi \restriction \alpha = id.$$

Moreover $\pi \restriction \text{TC}(\mathfrak{M}) \cup \text{TC}(\mathfrak{N}) \cup \{\mathfrak{M}, \mathfrak{N}\} = id$. It is easy to see then from the absoluteness of our process that

$$\pi^{-1}(I_\theta) = I_\alpha,$$

and

$$\pi^{-1}(J_\theta) = J_\alpha.$$

Subclaim. $\pi \restriction M_\alpha = i_{\alpha,\infty}$ and $\pi \restriction N_\alpha = j_{\alpha,\infty}$.

Proof: [Proof of the Subclaim] Note $M_\alpha = M_\infty^{I_\alpha} = \pi^-(M_\infty^{I_\theta})$, and for all $\beta < \alpha$, $i_{\alpha\beta} = \pi^{-1}(i_{\beta,\infty})$. So if $x \in M_\alpha$, let $x = i_{\beta\alpha}(\bar{x})$ where $\beta < \alpha$, and then

$$\begin{aligned}\pi(x) &= \pi(i_{\beta\alpha}(\bar{x})) \\ &= \pi(i_{\beta\alpha})(\bar{x}) \\ &= i_{\beta\infty}(\bar{x}) \\ &= i_{\alpha,\infty}(x).\end{aligned}$$

The same proof shows that $\pi \restriction N_\alpha = j_{\alpha,\infty}$. □

So we have $\alpha = \text{crit}(\pi) = \text{crit}(i_{\alpha,\infty}) = \text{crit}(j_{\alpha,\infty})$. Now notice the critical points used in I_θ are strictly increasing, and therefore

$$\alpha = \text{crit}(i_{\alpha,\alpha+1}) < \text{crit}(i_{\alpha+1,\infty}).$$

Similarly,

$$\alpha = \text{crit}(j_{\alpha,\alpha+1}) < \text{crit}(j_{\alpha+1,\infty}).$$

This is the crucial use of coherence! By coherence, $\mathfrak{M}_{\alpha+1}$ and $\mathfrak{N}_{\alpha+1}$ agree on all measures with crit $\leq \alpha$, and this agreement will never be disturbed later.

Exercise 15: Provide the details here.

So we had a diagreement of type (b) at α. Let U and W be the disagreeing ultrafilters, i.e. $I_{\alpha+1} = I_\alpha{}^\frown\langle U\rangle$ and $J_{\alpha+1} = J_\alpha{}^\frown\langle W\rangle$. We must have a set $A \subseteq \alpha$ such that $A \in M_\alpha \cap \mathfrak{N}_\alpha$ and $A \in U \triangle W$. Say $A \in U$ and $A \notin W$, then

$$\alpha \in i_{\alpha,\alpha+1}(A) = i_U^{M_\alpha}(A),$$

and

$$\alpha \notin j_{\alpha,\alpha+1}(A) = i_W^{\mathfrak{N}_\alpha}(A).$$

So

$$\alpha \in i_{\alpha,\infty}(A),$$

and

$$\alpha \notin j_{\alpha,\infty}(A),$$

because neither $i_{\alpha+1,\infty}$ nor $j_{\alpha+1,\infty}$ moves α. Since $i_{\alpha,\infty}(A) = \pi(A) = j_{\alpha,\infty}(A)$, we have a contradiction.

This finish the proof of the **Claim**. □

This finish the proof of Lemma 4.8. □

It is easy to see that linear iterability is π_1-definable in the language of set theory. So we get:

Theorem 4.10: *Assume $V = L[\mathcal{U}]$, where $\mathcal{U}$ is a coherent sequence of nufs. Then* **CH** *holds and $\mathbb{R}$ admits a Δ_2^{HC} well-order.*

Proof: Let $<^{\mathcal{U}}$ be "the" order of construction in $L[\mathcal{U}]$. (It is unique up to how we order the formulae. Fix one recursive ordering of formulae.) Then for $x, y \in \mathbb{R}$

$$\begin{aligned} x <^{\mathcal{U}} y \text{ iff } &\exists(M, \mathcal{W}) \in \mathrm{HC}((M, \mathcal{W}) \text{ is a measures-premouse} \\ &\wedge(M, \mathcal{W}) \text{ is } \omega_1\text{-iterable } \wedge M \models x <^{\mathcal{W}} y). \end{aligned} \qquad (*)$$

The ($\Longrightarrow$) direction comes from Löwenheim-Skolem: if $x <^{\mathcal{U}} y$, we have some α such that

$$(L_\alpha[\mathcal{U}], \in, \mathcal{U}) \models \mathbf{ZFC}^- + x <^{\mathcal{U}} y,$$

and then we can take $(M, \mathcal{U})$ to be the transitive collapse of a countable elementary submodel of $(L_\alpha[\mathcal{U}], \in, \mathcal{U})$. For the ($\Longleftarrow$) direction of $(*)$, suppose $(M, \mathcal{W})$ is as on the right hand side, but $y \leq^{\mathcal{U}} x$. Let $(L_\alpha[\mathcal{U}], \in, \mathcal{U}) \models y \leq^{\mathcal{U}} x$. Now bt Lemma 4.8, we can compare $\mathfrak{M} = (M, W)$ with $\mathfrak{N} = (N, \mathcal{U})$. The comparison maps do not move x and y, since they are reals. This is a contradiction.

The same proof shows that the order-type of $<^{\mathcal{U}} \restriction \mathbb{R}$ is ω_1, so **CH** holds. □

Exercise 16:

(a) Provide the details of the proof that $\mathfrak{M}$ and $\mathfrak{N}$ cannot be compared.
(b) Provide the details of the proof that **CH** holds in $L[\mathcal{U}]$.

Exercise 17: Assume $V = L[\mathcal{U}]$, where $\mathcal{U}$ is a coherent sequence of nufs.

(a) Show that $2^\alpha = \alpha^+$, except possibly when $\exists\kappa(\kappa < \alpha \wedge \alpha^+ \leq o^{\mathcal{U}}(\kappa))$.
(b) Show that if $o\mathcal{U}(\kappa) \leq \kappa^{++}$ for all κ, then **GCH** holds.
(c) Show that $o^{\mathcal{U}}(\kappa) \leq \kappa^{++}$ for all κ.
(d) (Harder) Show that **GCH** holds. (The hard case is $2^\alpha = \alpha^+$ when $\alpha = \kappa^+$ and $o^{\mathcal{U}}(\kappa) = \kappa^{++}$.)

The proof of Theorem 3.14 also shows that every real in a measure-mouse is ordinal definable in a simply way:

Exercise 18: Let $x \in \mathbb{R} \cap M$, where $(M,\mathcal{U})$ is a measure-mouse. Show that x is $\Delta_2^{\mathrm{HC}}(\{\alpha\})$, for some $\alpha < \omega_1$.

As corollaries to Theorem 4.10 and its proof, we get limitations on the consequences of (many) measurable cardinals we draw in Lecture 3. For example

Corollary 4.11: *Suppose $\exists j : V \to M(\mathrm{crit}(j) = \kappa \wedge V_{\kappa+2} \subseteq M)$. Then there is a model of* $\mathbf{ZFC} + \exists\alpha(o(\alpha) = \alpha^{++})$ *in which not all Δ_2^{HC} sets are Lebesgue measurable.*

Similarly, Martin-Solovay's generic absoluteness result (Theorem 3.14) does not extend to Σ_3^{HC}, even assuming many measurables. For let

$$\varphi = \forall x \in \mathbb{R} \exists \mathfrak{M}(\mathfrak{M} \text{ is a measures-mouse } \wedge x \in \mathfrak{M}).$$

One can calculate that φ is Π_3.

Exercise 19: Let $L[\mathcal{U}] \models$ "$\mathcal{U}$ is coherent", then

(1) $\mathrm{HC}^{L[\mathcal{U}]} \models \varphi$,
(2) whenever x is Cohen-generic$/L[\mathcal{U}]$, $\mathrm{HC}^{L[\mathcal{U}][x]} \models \neg\varphi$,
(3) φ is Π_3.

Finally, the correctness of models of the form $L[\mathcal{U}]$, $\mathcal{U}$ is coherent, is limited. For suppose there is $j : V \to M$ with $V_{\mathrm{crit}(j)+2} \subseteq M$. Then there is a Σ_2 sentence ψ such that

$$\mathrm{HC} \models \psi,$$

but whenever $\mathfrak{M}$ is a measures-mouse

$$\mathrm{HC}^{\mathfrak{M}} \not\models \psi.$$

Exercise 20: (For the student who knows more inner model theory.) Prove this.

5. Extenders

Given $j : M \to N$ elementary, it may not be the case that $N = \mathrm{Hull}^N(\mathrm{ran}(j) \cup \{\kappa\})$, for $\kappa = \mathrm{crit}(j)$. That is we may not have $N \cong \mathrm{Ult}(M,\mathcal{U})$. In order to capture j in general, we need a certain system of ultrafilters, called an extender.

Definition 5.1: $[X]^n = \{a \subseteq X \mid |a| = n\}$, and $[x]^{<\omega} = \bigcup_{n<\omega}[X]^n$. For $a \subseteq \mathrm{Ord}$ with $|a| = n$, we write $a_i = i^{th}$ element of a in its increasing enumeration.

Now suppose $j : M \to N$ with $M \models \mathbf{ZFC}^-$ transitive. Suppose $\lambda \subseteq \text{wfp}(N)$, and $\lambda \leq^N j(\kappa)$. For $a \subseteq [\lambda]^{<\omega}$ and $X \subseteq [\kappa]^{|a|}$ with $X \in M$, we put

$$X \in E_a \quad \text{iff} \quad a \in j(X).$$

Definition 5.2: $E_j^\lambda = \{(a, X) \mid X \in E_a\}$ is the (κ, λ)-**extender** derived from j.

Remark 5.3: The restriction $\lambda \leq j(\kappa)$ is not really needed. Extenders satisfying it are called "short extender", but since we have no use here for "long" extenders, we have dropped the qualifier "short" in these lectures.

Clearly, $E_{\{\kappa\}} = U_j$. By allowing typical objects beyond κ to generate measures, however, we may be able to capture more of j.

As before, we have

(1) E_a is an ultrafilter on $\mathcal{P}([\kappa]^{|a|})^M$, and non-principal iff $a \not\subseteq \kappa$,
(2) E_a is M-κ-complete. We can form $\text{Ult}(M, E_a)$, and we have

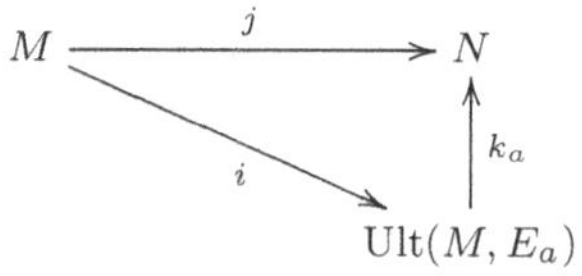

Fig. 10.

commutes, where $k_a([f]) = j(f)(a)$. (So $k_a([id]) = a$.) the range of k_a is $\text{Hull}^N(\text{ran}(j) \cap a)$.

If $a \subseteq b$, there is a natural map

$$i_{ab}(x) = k_b^{-1}(k_a(x)).$$

Since $[\lambda]^{<\omega}$ is directed under inclusion, and the maps commute (i.e. $i_{ac} = i_{bc} \circ i_{ab}$ if $a \subseteq b \subseteq c$), we can set

$$\text{Ult}(M, E) = \text{direct limit of Ult}(M, E_a)\text{'s under } i_{ab}\text{'s}.$$

We can piece the k_a's together into

$$k : \text{Ult}(M, E) \to N$$

given by $k(i_{a,\infty}(x)) = k_a(x)$, for $i_{a,\infty} : \text{Ult}(M, E_a) \to \text{Ult}(M, E)$ the direct limit map. The following commutative diagram summarizes things:

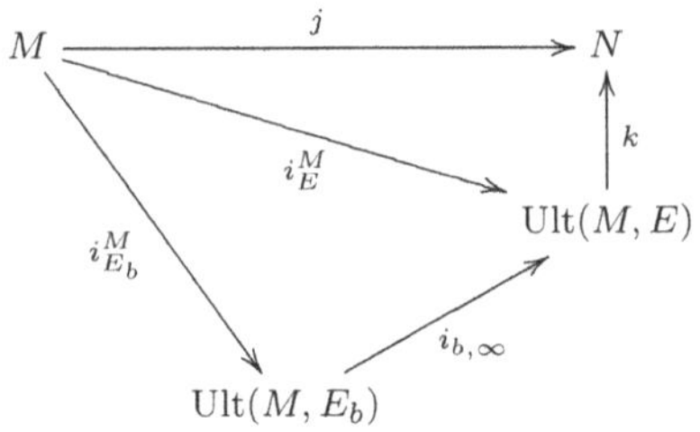

Fig. 11.

Note $\mathrm{ran}(k) = \mathrm{Hull}^N(ran(j) \cup \lambda)$, so that $k \restriction \lambda =$ identity.

Properties of E_j^λ

For $a \subseteq b$ and $b \in [\lambda]^{<\omega}$, and $X \subseteq [\kappa]^{|a|}$, we think of X as a predicate of $|a|$-tuples, and let X^{ab} be the result of "adding dummy variables" corresponding to ordinals in $b - a$. That is, for

$$a = \{b_{i_1}, \ldots, b_{i_n}\} \text{ with } i_1 < \cdots < i_n,$$

we put

$$X^{ab} = \Big\{u \in [\kappa]^{|b|} \mid \{u_{i_1}, \ldots, u_{i_n}\} \in X\Big\}.$$

Similarly, if $f : [\kappa]^{|a|} \to M$,

$$f^{ab}(u) = f(\{u_{i_1}, \ldots, u_{i_n}\}), \text{ for } u \in [\kappa]^{|b|}.$$

We then have the following properties of $E = E_j^\lambda$, where $j : M \to N$ with $crit(j) = \kappa$ and $\lambda <^N j(\kappa)$;

(1) Each E_a is an M-κ-complete ultrafilter on $\mathcal{P}([\kappa]^{|a|})^M$,
(2) (Compatibility) If $a \subseteq b$, then $\forall X \in M$

$$X \in E_a \quad \text{iff} \quad x^{ab} \in E_b,$$

(3) (M-normality) If $f \in M$ with $\mathrm{dom}(f) = [\kappa]^{|a|}$, and

$$f(u) < u_i, \text{ for } E_a \text{ a.e. } u,$$

then there is a $\xi < a_i$ such that letting $\xi = (a \cup \{\xi\})_k$,

$$f^{a, a\cup\{\xi\}}(u) = u_k, \text{ for } E_{a\cup\{\xi\}} \text{ a.e. } u.$$

If in addition $\mathcal{P}(\kappa)^M = \mathcal{P}(\kappa)^N$

(4) (M-amenability) If $a \in [\lambda]^{<\omega}$ and $a \subseteq \mathcal{P}([\kappa]^{|a|})$ with $a \in M$ and $M \models |a| \leq k$, then $E_a \cap a \in M$.

Definition 5.4: Given $M \models \mathbf{ZFC}^-$, transitive, and we call a system $E = \langle E_a \mid a \in [\lambda]^{<\omega}$ satisfying (1)-(4) above a **(κ, λ)-pre-extender over** M (or just an **M-pre-extender**). We write $\kappa = \text{crit}(E)$ and $\lambda = \text{lh}(E)$.

Notice that in Definition 5.4, we have thrown away j and N. In particular, if $Q \models \mathbf{ZFC}^-$ is transitive, and $\mathcal{P}(\kappa)^Q = \mathcal{P}(\kappa)^M$, then E is an M-pre-extender iff E is a Q-pre-extender.

If E is a (κ, λ)-pre-extender over M, then we define $\text{Ult}(M, E)$ as follows:

The elements are equivalence classes $[a, f]_E^M$, where for $f, g \in M$ with domains $[\kappa]^{|a|}$ and $[\kappa]^{|b|}$,

$$\langle a, f \rangle \sim \langle b, g \rangle \text{ iff for } E_{a \cup b} \text{ a.e. } u,\ f^{a, a \cup b}(u) = g^{b, a \cup b}(u),$$

and

$$[a, f]_E^M \widetilde{\in} [b, g]_E^M \text{ iff for } E_{a \cup b} \text{ a.e. } u,\ f^{a, a \cup b}(u) \in g^{b, a \cup b}(u).$$

Then we set

$$\text{Ult}(M, E) = (\{[a, f]_E^M \mid a \in [\lambda]^{<\omega} \wedge f \in M\}, \widetilde{\in}).$$

Let also $i_E^M : M \to \text{Ult}(M, E)$ be given by

$$i_E^M(x) = [\{0\}, \lambda u.x],$$

we have

(1) Łoś Theorem: given $\langle a_0, f_0 \rangle, \ldots, \langle a_n, f_n \rangle$ and $\varphi(v_0, \ldots, v_n)$ and letting $b = \bigcup_{i \leq n} a_i$,

$$\text{Ult}(M, E) \models \varphi[[a_0, f_0]_E^M, \ldots, [a_n, f_n]_E^M] \text{ iff for } E_b \text{ a.e.}$$
$$u\ M \models \varphi[f_0^{a_0, b}(u), \ldots, f_n^{a_n, b}(u)].$$

(2) i_E^M is elementary,
(3) $\text{crit}(i_E^M) = \kappa$,
(4) letting $id(u) = u$ for all $u \in [\kappa]^{|a|}$,

$$[a, id]_E^M = a,$$

and

$$[a, f]_E^M = i_E^M(f)(a).$$

(5) $X \in E_a$ iff $a \in i_E^M(X)$.

Exercise 21: Prove (1)-(5).

The Ult(M, E) is the Skolem-closure of ran$(i_E^M) \cup \lambda$ inside Ult(M, E), and E is the (κ, λ)-pre-extender derived from Ult(M, E). By amenability, $\mathcal{P}(\kappa)^M = \mathcal{P}(\kappa)^{\text{Ult}(M,E)}$, the closure of Ult$(M, E)$ under sequences is given by:

Lemma 5.5: *Let E be a (κ, λ)-pre-extender over M, and $\alpha \leq \kappa$. Suppose M is closed under α-sequences, and ${}^\alpha\lambda \subseteq$ Ult(M, E). Then* Ult(M, E) *is closed under α-sequences.*

Proof: Let $[a_\beta, f_\beta] \in$ Ult(M, E) for all $\beta < \alpha$. Let $i = i_E^M$. Then

$$\langle i(f_\beta) \mid \beta < \alpha \rangle = i(\langle f_\beta \mid \beta < \alpha \rangle) \restriction \alpha \in M$$

and $\langle a_\beta \mid \beta < \alpha \rangle \in M$ as $\lambda_\alpha \subseteq M$. Thus $\langle i(f_\beta)(a_\beta) \mid \beta < \alpha \rangle \in M$. □

Exercise 22: Let κ be measurable. Show there is a (κ, λ)-extender over V such that Ult(V, E) is not closed under ω-sequences.

We have $\lambda \subseteq$ wfp(Ult(M, E)), essentially by normality. How to guarantee Ult(M, E) is fully well-founded?

Definition 5.6: Let E be a (κ, λ)-pre-extender over M. We say E is **ω-complete** iff whenever $X_i \in E_a$ for all $i < \omega$, then there is an $f : \bigcup_{i<\omega} a_i \to \kappa$ such that

$$f\text{''}a_i \in X_i$$

for all $i < \omega$.

One sometimes calls f a "fiber" for $\langle (a_i, X_i) \mid i \in \omega \rangle$.

Lemma 5.7: *Let E be an ω-complete extender over M. Then* Ult(M, E) *is well-founded.*

Proof: Suppose $[a_{i+1}, g_{i+1}] \in [a_i, g_i]$ for all i. By meeting the right measure one sets, we can find a fiber f such that $g_{i+1}(f\text{''}a_{i+1}) \in g_i(f\text{''}a_i)$ for all i. This is a contradiction. □

Exercise 23: Let E be an ω-complete pre-extender over M, and $E \in M$. Let $\pi : N \to M$ be elementary, with N countable transitive, and $\pi(F) = E$. Show that there is a σ such that

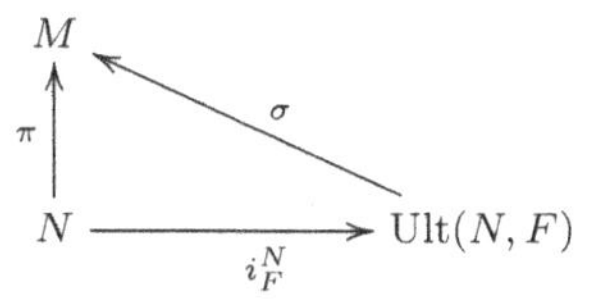

Fig. 12.

commutes. (That is, countable fragments of $\mathrm{Ult}(M, E)$ can be realized back in M.)

Exercise 24: (a) Let E be an pre-extender over V. Then $\mathrm{Ult}(V, E)$ is well-founded iff E is ω-complete.

(b) There is a pre-extender over V such that $\mathrm{Ult}(V, E)$ is ill-founded.

Definition 5.8: E is a (κ, λ)**-extender over** M iff E is a (κ, λ)-pre-extender over M, and $\mathrm{Ult}(M, E)$ is well-founded.

In contrast pre-extender-hood, there are M,Q, and E such that E is a (κ, λ)-extender over M, and $\mathcal{P}(\kappa)^M = \mathcal{P}(\kappa)^Q$, but E is not a (κ, λ)-extender over Q, because $\mathrm{Ult}(Q, E)$ is ill-founded. If $j : M \to N$ where N is well-founded, then E_j is indeed an extender over M:

Lemma 5.9: *Let $j : M \to N$ where M is transitive, and E be the (κ, λ)-extender derived from j, Then the diagram*

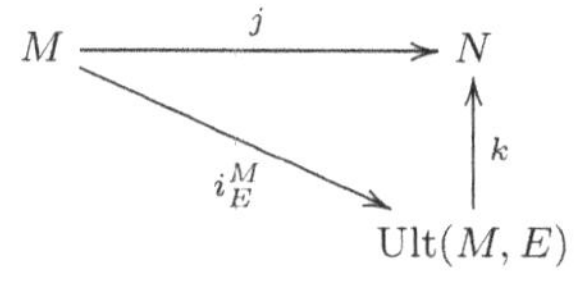

Fig. 13.

commutes, where $k([a, f]_E^M) = j(f)(a)$. Moreover, $k \restriction \lambda =$ identity.

Of course, we began this section by essentially proving Lemma 4.8. But there we had defined $\mathrm{Ult}(M, E)$ in a slightly different way. You can think of Lemma 4.8 as saying that the two constructions give the same $\mathrm{Ult}(M, E)$. We leave this proof as an informal exercise.

It follows that if N is well-founded, so is $\mathrm{Ult}(M, E)$.

Computing Large Cardinal Strength

If U is a nuf on κ, then $V_{\kappa+2} \not\subseteq \mathrm{Ult}(V, U)$. With extenders, more is possible.

Lemma 5.10: *Let $j : V \to N$, $\mathrm{crit}(j) = \kappa$, and $\kappa < \lambda \leq j(\kappa)$. Suppose $V_\alpha \subset N$, and $|V_\alpha|^+ < \lambda$ Let E be the (κ, λ)-extender derived from j, then $V_\alpha \subseteq \mathrm{Ult}(V, E)$.*

Proof: Let $k : \mathrm{Ult}(V, E) \to N$ be the factor map, as in Lemma 4.8. Let $\beta = |V_\alpha|$, so $\beta < \lambda$. Let $(\beta, R) \cong (V_\alpha, \in)$. Since $\beta, \alpha \in \mathrm{ran}(k)$, we can pick $R \in \mathrm{ran}(k)$. But then $k^{-1}(R) = R$, as $k \restriction \beta = id$. Since $R \in \mathrm{Ult}(V, E)$, $V_\alpha \subseteq \mathrm{Ult}(V, E)$. □

Definition 5.11: Let E be a V-extender; then $\mathrm{strength}(E) =$ largest α such that $V_\alpha \subseteq \mathrm{Ult}(V, E)$.

Exercise 25: Let E be a V-extender; then $E \notin \mathrm{Ult}(V, E)$, and therefore $\mathrm{strength}(E) \leq \mathrm{lh}(E)$.

Corollary 5.12: *Let $: V \to N$ where N is transitive, $\kappa = \mathrm{crit}(j)$, $\kappa < \lambda \leq j(\kappa)$. Suppose λ is inaccessible, and $V_\lambda \subseteq N$. Let E be the (κ, λ)-extender derived from j; then* $\mathrm{strength}(E) = \lambda$.

Definition 5.13: A V-extender E is **nice** iff $\mathrm{strength}(E) = \mathrm{lh}(E)$, and $\mathrm{strength}(E)$ is strongly inaccessible.

In the sequel, we shall use extenders (and iteration trees built from them) to extend the results of Martin-Solovay on correctness and generic absoluteness from Lecture 2. For these applications, nice extenders suffice. On the other hand, one can certainly not skip over the non-nice extenders in the bottom-up analysis of inner model theory.

We conclude this lecture with some simple lemmas on capturing large cardinal properties via extenders.

Definition 5.14:

(a) κ is β-strong iff $\exists j : V \to M$ (M transitive $\wedge \text{crit}(j) = \kappa \wedge V_\beta \subseteq M$).
(b) κ is superstrong iff $\exists j : V \to N$ (M transitive $\wedge \text{crit}(j) = \kappa \wedge V_{j(\kappa)} \subseteq M$).
(c) κ is λ-supercompact iff $\exists j : V \to M$ (M transitive $\wedge \text{crit}(j) = \kappa \wedge {}^\lambda M \subseteq M$).

Proposition 5.15: *If κ is 2^κ-supercompact, then κ is a limit of superstrong cardinals.*

Proof: Let $j : V \to M$, $\text{crit}(j) = \kappa$, and ${}^{2^\kappa}M \subseteq M$. So $j \restriction V_{\kappa+1} \in M$. Let E be the $(\kappa, j(\kappa))$ extender derived from j. We then have $E \in M$.

Exercise 26: $M \models \kappa$ is superstrong, as witnessed by E.

The exercise easily yields the proposition. □

Exercise 27: Let κ be superstrong. Show $\exists \alpha < \kappa \forall \beta < \kappa$ (α is β-strong).

Definition 5.16: Let $\kappa < \delta$ and $A \subseteq V_\delta$; then κ **is A-reflecting in** δ iff

$$\forall \beta < \delta \exists j : V \to M(\text{crit}(j) = \kappa \wedge V_\beta \subseteq M \wedge j(A) \cap V_\beta = A \cap V_\beta).$$

Definition 5.17: δ is **Woodin** iff $\forall A \subseteq V_\delta \exists \kappa < \delta$ (κ is A-reflecting in δ).

Proposition 5.18: *Suppose δ is Woodin. Then δ is strongly inaccessible, and there are arbitrarily large $\kappa < \delta$ such that $\forall \beta < \delta$ (κ is β-strong). The least Woodin cardinal is not Mahlo.*

We leave the easy proof to the reader.

The following notations is quite useful.

Definition 5.19: Let E be a (κ, λ)-pre-extender, and $X \subseteq \lambda$; then $E \restriction X = \{(a, Y) \mid a \in [X]^{<\omega} \wedge Y \in E_a\}$.

Mostly we use this when $X = \eta \leq \lambda$. If $\kappa < \eta \leq \lambda$, then $E \restriction \eta$ is itself a (κ, λ)-pre-extender.

Remark 5.20: We do not have to say "pre-extender over M", because pre-extender hood only depends on $\mathcal{P}(\kappa)^M$, which is determined by E itself. ($\mathcal{P}(\kappa)^M = E_{\{\kappa\}} \cup \{\kappa - A \mid A \in E_{\{\kappa\}}\}$.)

Lemma 5.21: *Let κ be superstrong; then κ is a Woodin limit of Woodin cardinals.*

Proof: Let $j : V \to M$ witness κ is superstrong. We show first κ is Woodin. So let $A \subseteq V_\kappa$.
Claim. $M \models \exists\alpha < j(\kappa)(\alpha$ is $j(A)$-reflecting in $j(\kappa))$.

Proof: [Proof of the Claim] Take $\alpha = \kappa$. Let $\kappa < \beta < j(\kappa)$. We may assume β is inaccessible in M. Let $E = E_j \restriction \beta$, so that $E \in M$. We have

$$A \cap V_\kappa = j(A) \cap V_\kappa,$$

so

$$j(A) \cap V^M_{j(\kappa)} = j(j(A)) \cap V^M_{j(\kappa)},$$

so

$$j(A) \cap V^M_\beta = i^M_E(j(A)) \cap V^M_\beta.$$

(To see the last line, note that

$$i^M_E(j(A)) \cap V^M_\beta = i^M_E(A) \cap V^M_\beta,$$

because $\beta < i^M_E(\kappa)$, and $i^M_E(A) \cap V^M_\beta = j(A) \cap V^M_\beta$, because $E = E_j \restriction \beta$.)
This gives the claim. □

Pulling the claim back to V, we get $V \models \exists\alpha < \kappa$ (α is A-reflecting in κ). Since A was arbitrary, κ is Woodin. But now it easily follows that

$$M \models \kappa \text{ is Woodin},$$

so

$$V \models \kappa \text{ is a limit of Woodins}.$$

We have then the following consistency strength hierarchy on those properties. (Note that κ measurable iff κ is $\kappa + 1$-strong.)

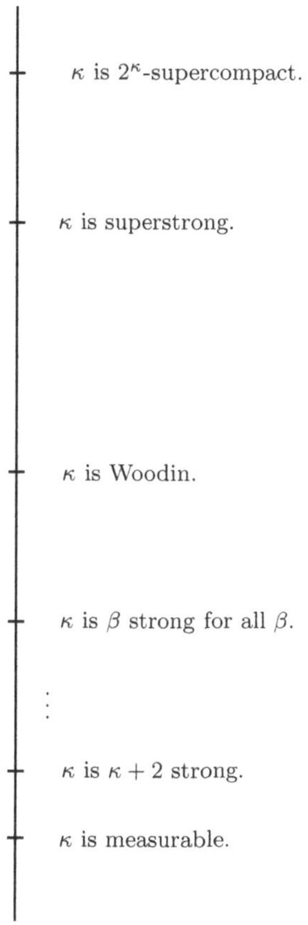

Fig. 14.

6. Linear Iteration via Extenders

Everything works much as it did with linear iteration via nufs, so we shall go quickly here.

Definition 6.1: An extender pair is an $(M, \mathcal{E})$ such that M is transitive, and $M \models \mathbf{ZFC}^- + \mathcal{E}$ is a set of V-extenders.

A linear iteration of $(M, \mathcal{E})$ is a sequence $\langle E_\alpha \mid \alpha < \beta \rangle$ determining M_α's and $j_{\alpha\gamma} : M_\alpha \to M_\gamma$, as before. $E_\alpha \in i_{0\alpha}(\mathcal{E})$, and $M_{\alpha+1} = \mathrm{Ult}(M_\alpha, E_\alpha)$.

Direct limits are taken at limit stages. M^I_∞ is the "last model" associated to the iteration I. $(M, \mathcal{E})$ is α-linearly iterable iff M^I_∞ is well-founded, for all I with $\mathrm{lh}(I) < \alpha$.

The following facts are proved just as they were for linear iterations of nuf-pairs.

(1) Let $(M, \mathcal{E})$ be an extender pair, $\pi : N \to M$ elementary, $\pi(\mathcal{F}) = \mathcal{E}$. Then if $(M, \mathcal{E})$ is α-linearly iterable, so is $(N, \mathcal{F})$. (Cf. Lemma 3.5.)
(2) Let $(M, \mathcal{E})$ be an extender pair. Then $(M, \mathcal{F})$ is ω_1-linearly iterable. (Cf. Lemma 3.6.)
(3) Let $(M, \mathcal{E})$ be an extender pair such that every $E \in \mathcal{E}$ is ω-complete. Then $(M, \mathcal{E})$ is linearly iterable. (Cf. Theorem 3.3, and Exercise 23.)
(4) Let $(M, \mathcal{E})$ be an extender pair such that $M \models$ **ZFC** and $\omega_1 \in M$. Then $(M, \mathcal{E})$ is linearly iterable. (Cf. Corollary 3.9.)

Applications

In Lecture 3, we showed that if there is a linearly iterable nuf pair $(M, \{U\})$ with "one measurable cardinal", then

(i) $\underset{\sim}{\Pi}^1_1$ determinacy holds.
(ii) $\underset{\sim}{\Sigma}^1_2$ sets are Lebesgue Measurable, etc.
(iii) M is $\underset{\sim}{\Sigma}^1_2$ correct.
(iv) M is Σ^1_3 correct in $M[G]$, when G is M-generic for $\mathbb{P} \in V^M_\kappa$, with $\kappa = \mathrm{crit}(U)$, and $M \models$ **ZFC**.

We showed that you cannot add I to any of the subscripts in (i)-(iv), in Lecture 4. This is true no matter how many measures your nuf pair $(M, \mathcal{E})$ is assumed to have. It remains true if we replace $(M, \mathcal{E})$ with an extender pair.

The basic reason is that linear iterability is $\Pi^{\mathrm{HC}}_1 = \Pi^1_2$. Thus for any sentence φ, the statement

ψ = "There is a countable, linearly iterable extender-pair $(M, \mathcal{E})$ such that $(M, \mathcal{E}) \models \varphi$".

(Think of φ as saying "There are extenders witnessing superstrongness", if you like.) If there are linearly iterable $(M, \mathcal{E}) \models \varphi$, then they cannot all be Σ^1_3 correct. For let $(M, \mathcal{E}) \models \varphi$ with $\mathrm{Ord} \cap M$ minimal; then $M \not\models \psi$.

This is not to say that the existence of linearly iterable extender pairs $(M, \mathcal{E})$ with "many extenders in $\mathcal{E}$" does not lead to strengthening of (i)-(iv) above. You just cannot go all the way to $\underset{\sim}{\Pi}^1_2$ in (i), or to $\underset{\sim}{\Sigma}^1_3$ in (ii) or (iii), or to Σ^1_4 in (iv). Instead, one needs to replace $\underset{\sim}{\Pi}^1_1$ in (i) by some level Γ of $\underset{\sim}{\Delta}^1_2$, and $\underset{\sim}{\Sigma}^1_2$ in (ii) and (iii) by a corresponding level $\circlearrowleft \Gamma$ of $\underset{\sim}{\Delta}^1_3$. (At the moment, we do not see how to extend (iv).) Such strengthenings of (i)-(iv) were developed by D.A. Martin, his students, and others in the 1970's and early 1980's.

Canonical Extender Models

If E and F are V-extenders, then again

$$E \lhd F \quad \text{iff} \quad E \in \mathrm{Ult}(V, F).$$

A coherent sequence of extenders will again be a sequence linearly ordered by $\lhd$, without leaving gaps. Here we face the complication that there certaninly E with $\eta < \mathrm{lh}(E)$ such that E is equivalent to $E \restriction \eta$. in that $\mathrm{Ult}(V, E) = \mathrm{Ult}(V, E \restriction \eta)$.

Exercise 28: If $\mathrm{crit}(E) < \gamma$ and $\gamma + 2 = \mathrm{lh}(E)$, then $\mathrm{Ult}(V, E \restriction (\gamma + 1) = \mathrm{Ult}(V, E)$.

So we can put the "same" extender on our sequence with different lengths. Some indexing convention is needed. Here is one from [6].

Definition 6.2: A coherent sequence of non-overlapping extenders is a function $\mathcal{E}$ with domain of the form $\{(\kappa, \beta) \mid \beta < o^{\mathcal{E}}(\kappa)\}$ such that

(1) If $o^{\mathcal{E}}(\kappa) > 0$, then $\forall \lambda < \kappa(o^{\mathcal{E}}(\lambda) < \kappa)$, and if $\beta < o^{\mathcal{E}}(\kappa)$, then
(2) $\mathcal{E}(\kappa, \beta)$ is a $(\kappa, \kappa + 1 + \beta)$ extender over V,
(3) $i_{\mathcal{E}(\kappa,\beta)}(\mathcal{E}) \restriction (\kappa + 1, 0) = \mathcal{E} \restriction (\kappa, \beta)$.

The "non-overlapping" part is clause (1). It guarantees that $\mathcal{E}$ is simple enough that iterability suffices for canonicity (e.g., for comparison). On the other hand, it prevents $L[\mathcal{E}]$ from satisfying more than "There is a strong cardinal". If we want a theory of canonical inner models with for example, Woodin cardinals, then drop clause (1) above, but at the same time we must generalize the notion of linear iterability.

Moving to extenders yields one simplification: the functions witnessing coherence are now trivial, as if $\gamma < \beta$, then $\gamma = [\{\gamma\}, id]_{\mathcal{E}(\kappa,\beta)}$. So we get

Proposition 6.3: *If $\mathcal{E}$ is a coherent sequence of non-overlapping extenders, then*

$$L[\mathcal{E}] \models \mathcal{E} \textit{ is a coherent sequence of non-overlapping extenders.}$$

Remark 6.4: $L[\mathcal{E}] = L[A]$, where $A = \{(\kappa, \beta, a, x) \mid (a, x) \in \mathcal{E}(\kappa, \beta)\}$.

Also, it becomes a little easier to show that large cardinal properties go down to $L[\mathcal{E}]$. For example,

Theorem 6.5: *Suppose there is a strong cardinal; then there is a proper class $\mathcal{E}$ such that*

$$L[\mathcal{E}] \models \mathcal{E} \text{ is a coherent sequence of non-overlapping extenders,}$$

and

$$L[\mathcal{E}] \models \mathcal{E} \text{ there is a strong cardinal.}$$

Proof: [Proof Sketch] Construct a maximal non-overlapping coherent sequence, defining $\mathcal{E}(\kappa, \beta)$ by induction on the lexicographic order on the (κ, β)'s. □

Exercise 29: Give a real proof of Theorem 6.5.

Definition 6.6: An **extenders-premouse** is a pair $(M, \mathcal{E})$ such that M is transitive and

$$M \models \mathbf{ZFC}^{-}\mathcal{E} \text{ is a coherent sequence of non-overlapping extenders.}$$

An **extenders-mouse** is a linearly iterable extenders-premouse.

The initial segment relation $\mathfrak{M} \trianglelefteq \mathfrak{N}$ on extenders-premice is defined just as before. We get as before:

Theorem 6.7: (Comparison Lemma) *Let $\mathfrak{M}$ and $\mathfrak{N}$ be set size extender-mice. Then there are linear iterations I and J of $\mathfrak{M}$ and $\mathfrak{N}$ such that*

$$\mathfrak{M}^I_\infty \trianglelefteq \mathfrak{N}^J_\infty \text{ or } \mathfrak{N}^J_\infty \trianglelefteq \mathfrak{M}^I_\infty.$$

Corollary 6.8: *Suppose $V = L[\mathcal{E}]$, where $\mathcal{E}$ is a coherent sequence of non-overlapping extenders. Then* **CH** *holds, and $\mathbb{R}$ admits a Δ^{HC}_2 well order.*

We get other corollary parallel to those in Lecture 4 as well. For example,

Corollary 6.9: Con(**ZFC**+ *"There is a strong cardinal."*) $\Longrightarrow$ Con(**ZFC**+ *"There is a strong cardinal"*+ *" $\mathbb{R}$ admits a Δ^{HC}_2 wellorder."*)

The Comparison Lemma, Theorem 6.7, is proved just as Lemma 4.8, the Comparison Lemma for measures-mice, was proved. I and J are constructed by iteration away the least disagreement.

7. Iteration Trees of Length ω

Suppose

$$M_0 \models E_0 \text{ is an extender and } \lambda = \text{strength}(E_0).$$

Set

$$M_1 = \text{Ult}(M_0, E),$$

and suppose

$$M_1 \models E_1 \text{ is an extender, and } \text{crit}(E_1) < \lambda.$$

If we were iterating in linear fashion as before, our next model would be $M_2 = \text{Ult}(M_1, E_1)$. But there is another possibility. Since $V_\lambda^{M_0} = V_\lambda^{M_1}$ and $\text{crit}(E_1) < \lambda$, E_1 is a pre-extender over M_0, and we could set

$$M_2 = \text{Ult}(M_0, E_1),$$

and continue iterating from there. (Assuming that M_2 is well-founded!) For example, one can show that for $\lambda_1 = \text{strength}(E_1)^{M_1}$, we have that $V_{\lambda_1}^{M_1} = V_{\lambda_1}^{M_2}$. (This is done below.) So if

$$M_2 \models E_2 \text{ is an extender}$$

and $\text{crit}(E_2) < \lambda_1$, then we could set

$$M_3 = \text{Ult}(M_1, E_2),$$

and if M_3 were well-founded, continue from there. Our picture so far is

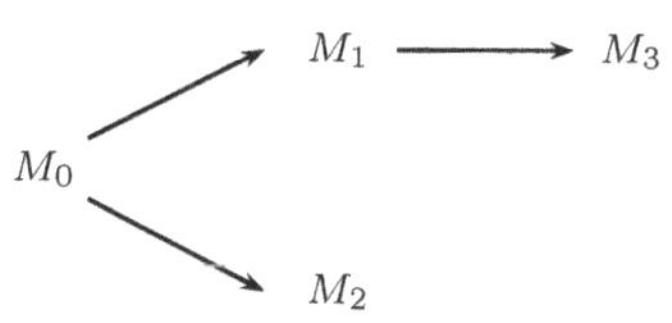

Fig. 15.

If we could find extenders with the right pattern of strengths and critical points, with all the relevant ultrapowers well-founded, we might be able to generate an "alternating chain":

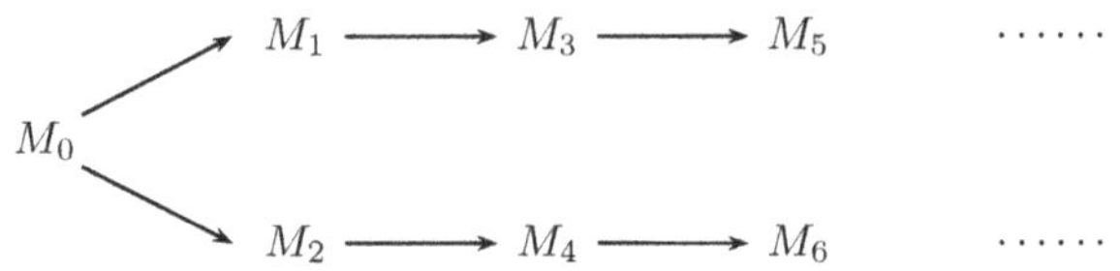

Fig. 16.

where

$$M_{n+1} = \mathrm{Ult}(M_{n-1}, E_n)$$

with

$$M_n \models E_n \text{ is an extender.}$$

It turns out that if $M_0 = V$, and $\exists\delta$ (δ is woodin), then there are indeed such alternating chains, and in fact, many other interesting and useful "iteration tree".

In this section, we shall restrict ourselves to models of full **ZFC**. Whenever we speak of $\mathrm{Ult}(M, E)$, we assume that M is transitive and $M \models$ **ZFC**, enen if this not explicitly stated.

Definition 7.1: For M, N transitive models of **ZFC**,

$$M \underset{\alpha}{\frown} N \quad \text{iff} \quad V_\alpha^M = V_\alpha^N.$$

The following helps prepare agreement-of-models in an iteration tree.

Lemma 7.2: *Let M and N be transitive models of* **ZFC**, *and $M \underset{\kappa+1}{\frown} N$. Suppose*

$$M \models E \textit{ is an extender}$$

and $\mathrm{crit}(E) = \kappa$. *Then*

(1) E is a pre-extender over N,
(2) $\mathrm{Ult}(, E) \underset{i_E(\kappa)+1}{\frown} \mathrm{Ult}(N, E)$, *and*
(3) $i_E^M \restriction V_{\kappa+1}^M = i_E^N \restriction V_{\kappa+1}^N$.

Proof: Sketch of proof Let $f : [\kappa]^{|a|} \to V_{\kappa+1}^M = V_{\kappa+1}^N$. Then $f \in M$ iff $f \in N$. This implies that the two ultrapowers agree on their common image of $V_{\kappa+1}^M = V_{\kappa+1}^N$. □

Exercise 30: Think through the details.

In order to simplify some points we shall for now only consider iteration trees formed using extenders which are nice in the model they are taken from. (Recall that E is nice iff $\mathrm{lh}(E) = \mathrm{strength}(E)$ is strongly inaccessible.) From Lemma 5.5, we get at once

Lemma 7.3: *Suppose $M \models E$ is a nice (κ, λ)-extender, and $M \underset{\kappa+1}{\frown} N$. LEt $\alpha \leq \kappa$, and suppose both M and N are closed under α-sequences, then* $\mathrm{Ult}(N, E)$ *is closed under α-sequences.*

In the situation of Lemma 7.3, if $\alpha \geq \omega$, then we can conclude that $\mathrm{Ult}(N, E)$ is well-founded.

Definition 7.4: Let $\alpha \leq \omega$. T is a **tree order on α** iff

(1) (α, T) is a partial order,
(2) $(nTm) \Longrightarrow n < m$, for all $m, n < \alpha$,
(3) $\{n \mid nTm\}$ is linearly ordered by T, for all $m < \alpha$, and
(4) $0\ Tn$ for all n such that $0 < n < \alpha$.

We write $\mathrm{pd}_T(n+1) =$ largest m such that $mT(n+1)$.

Definition 7.5: Let $\alpha \leq \omega$. A **nice iteration tree of length α on M** is a pair $\mathcal{T} = \langle T, \langle (M_n, E_n) \mid n < \alpha \rangle \rangle$ such that for all $n, m < \alpha$

(1) T is a tree order on α,
(2) $M_0 = M$,
(3) $M_n \models E_n$ is a nice extender,
(4) $n < m \Longrightarrow \mathrm{lh}(E_n) < \mathrm{lh}(E_m)$,
(5) if $n + 1 < \alpha$, then $M_{n+1} = \mathrm{Ult}(M_k, E_n)$, where $k =$ least i such that $M_i \underset{\mathrm{crit}(E_n)+1}{\frown} M_n$, moreover $\mathrm{pd}_T(n+1) = k$ in this case.

Notation 1: For $\mathcal{T}$ as above, we set $\mathrm{lh}(\mathcal{T}) = \alpha$. If $\mathcal{T}$ is an iteration tree, we write $M_n^{\mathcal{T}}$ and $E_n^{\mathcal{T}}$ for the models an extenders of $\mathcal{T}$. Note that if T is the associated tree order, then there are canonical embeddings

$$i_{nm}^{\mathcal{T}} : M_n^{\mathcal{T}} \to M_m^{\mathcal{T}} \quad (\text{for } nTm)$$

between the models earlier on a given branch and thise later. (Here $i_{k,m} : M_k \to \mathrm{Ult}(M_k, E_{m-1})$ is the canonical ultrapower embedding if $k = \mathrm{pd}_T(m)$, and $i_{k,m} = i_{\mathrm{pd}_T(m),m} \circ i_{k,\mathrm{pd}_T(m)}$ otherwise.)

The following lemma records the agreement between models in a nice iteration tree.

Lemma 7.6: *Let $\langle T, \langle (M_n, E_n) \mid n < \alpha \rangle\rangle$ be a nice iteration tree. Let $k \leq n$; then*

(a) $M_k \underset{\mathrm{lh}(E_k)}{\frown} M_n$, but
(b) it is not the case that $M_k \underset{\mathrm{lh}(E_k)+1}{\frown} M_n$, if $k < n$.

Moreover, for any n, $\mathrm{pd}_T(n+1)$ is the least i such that $\mathrm{crit}(E_n) < \mathrm{lh}(E_i)$.

Proof: We prove (a). Fix k. The proof is by induction on n. The case $n = k$ is clear. Not let $\lambda = \mathrm{lh}(E_k)$. We have

$$M_n \underset{\lambda}{\frown} M_k$$

by induction, and

$$M_n \underset{i_{e_n}(\kappa)+1}{\frown} M_{n+1}$$

by Lemma 7.2, where $\kappa = \mathrm{crit}(E_n)$. But

$$\mathrm{lh}(E_k) \leq \mathrm{lh}(E_n) \leq i_{E_n}(\kappa),$$

by clause (4) of Definition 7.5, and the fact that we use only short extenders. Thus $M_{n+1} \underset{\lambda}{\frown} M_k$, completing the induction step.

The rest is an exercise. □

Exercise 31. Complete the proof of Lemma 7.6.

Remark 7.7: Our nice iteration trees are fairly special in several ways. It is not necessary to use only nice extenders. It is not necessary that the strengths of the extenders be increasing, as in clause (4) of Definition 7.5. It is not necessary that $\mathrm{pd}_T(n+1)$ be the least i such that $M_i \underset{\mathrm{crit}(E_n)+1}{\frown} M_n$, it need only be some such i. A more general notion of iteration tree is needed in many contexts. The present restrictions make several points cleaner. In particular, Lemma 7.6 has the simple statement above.

Here is a diagram illustrating Lemma 7.6. In the diagram, $\mathrm{crit}(E_i) = \kappa_i$ and $\mathrm{strength}^{M_i}(E_i) = \lambda_i$.

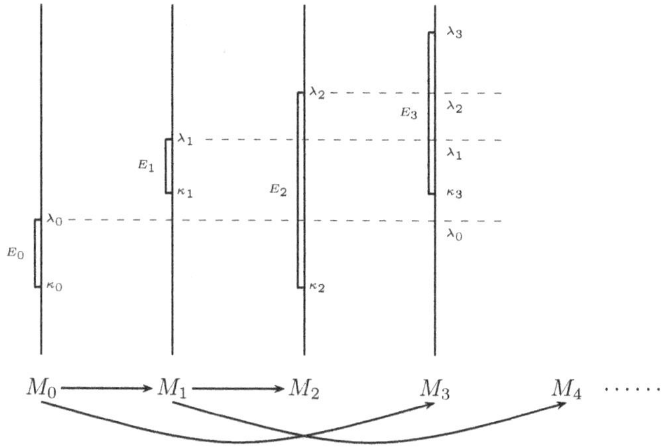

Fig. 17.

Another picture of the same iteration tree:

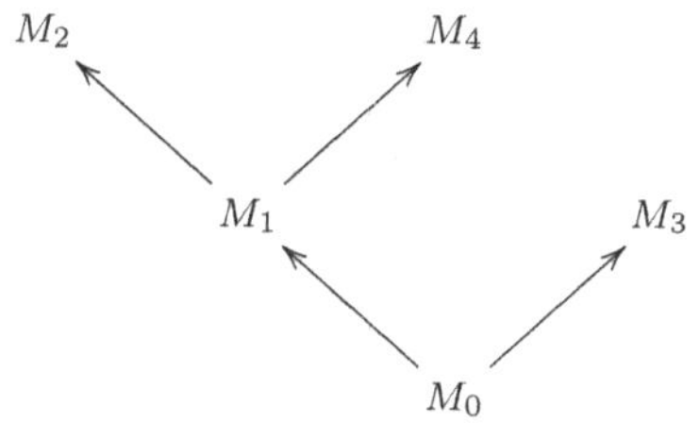

Fig. 18.

As the first picture shows, the process of determing and iteration tree is linear: M_{n+1} is determined by $\langle (M_i, E_i) \mid i \leq n\rangle$, and then we are free to choose E_{n+1} in order to continue. As the second picture shows, the embeddings between models may fall into a non-linear structure.

If $\mathcal{T}$ has length $n+1$, then we set

$$M_\infty^{\mathcal{T}} = \mathrm{Ult}(M_k, E_n^{\mathcal{T}}), \text{ where } k = \text{the least } i \text{ such that } \mathrm{crit}(E_n^{\mathcal{T}}) < \mathrm{lh}(E_i^{\mathcal{T}}),$$

and we call $M_\infty^{\mathcal{T}}$ be the **last model** of $\mathcal{T}$. Nothing forces it to be well-founded, but if we are to continue from $\mathcal{T}$, it had better be! In this connection, we have immediately from Lemma 7.3:

Lemma 7.8: *Let $\mathcal{T}$ be a nice iteration tree of length $\leq \omega$, and $\eta < \inf\{\mathrm{crit}(E_i^{\mathcal{T}}) \mid i+1 < \mathrm{lh}(\mathcal{T})\}$. Suppose $M_0^{\mathcal{T}}$ is closed under η-sequences. Then*

(a) $\forall n < \mathrm{lh}(\mathcal{T})$ ($M_n^{\mathcal{T}}$ is closed under η-sequences), and
(b) if $\mathrm{lh}(\mathcal{T}) < \omega$, then $M_\infty^{\mathcal{T}}$ is closed under η-sequences,
(c) if $\mathrm{lh}(\mathcal{T}) < \omega$, then $M_\infty^{\mathcal{T}}$ is well-founded.

Exercise 32: Prove Lemma 7.8. (It is easy.)

How do we continue an iteration tree of length ω?

Definition 7.9: Let $\mathcal{T} = \langle T, \langle M_n, E_n \mid n < \omega\rangle\rangle$ be an iteration tree, and let b be a branch of T. Then

$$M_b^{\mathcal{T}} = \mathrm{dir}\lim_{k\in b} M_k^{\mathcal{T}}$$

under the $i_{k,l}^{\mathcal{T}}$ for $k, l \in b$. We say b **is well-founded** iff $M_b^{\mathcal{T}}$ is well-founded.

What we would like, in order to continue from $\mathcal{T}$, is a confinal-in-ω well-founded branch of $\mathcal{T}$. If $M_0^{\mathcal{T}} = V$, then we can find such a branch:

Theorem 7.10: *Let $\mathcal{T}$ be a nice iteration tree on V, with $\mathrm{lh}(\mathcal{T}) = \omega$. Then there is a confinal-in-ω well-founded branch of $\mathcal{T}$.*

Remark 7.11: (1) So for example, the tree

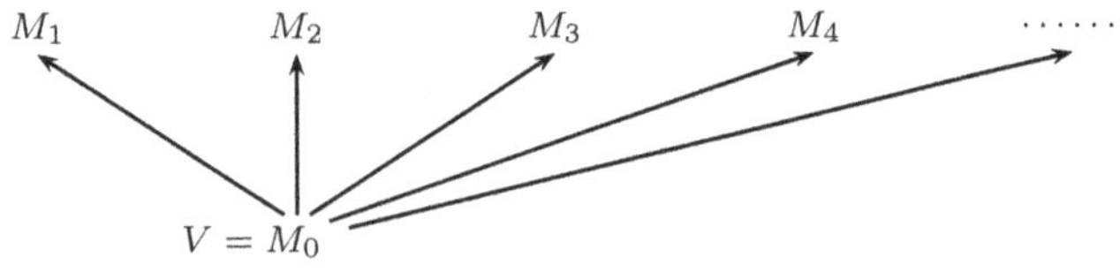

Fig. 19.

is impossible. In other words, the Mitchell order $\triangleleft$ on nice extenders is well-founded. (In fact, it is well-founded on arbitrary short extenders.)

(2) More generally, if $\mathcal{T} = \langle T, \cdots \rangle$ is a nice iteration tree on V of length ω, then T has an infinite branch. Andretta ([1]) has shown that this is the only restriction on T, provide there is a Woodin cardinal. That is, if there is a Woodin cardinal, and T is a tree order on ω having an infinite branch, then there is a nice iteration treeon V whose tree order is T.

By Theorem 7.10, if

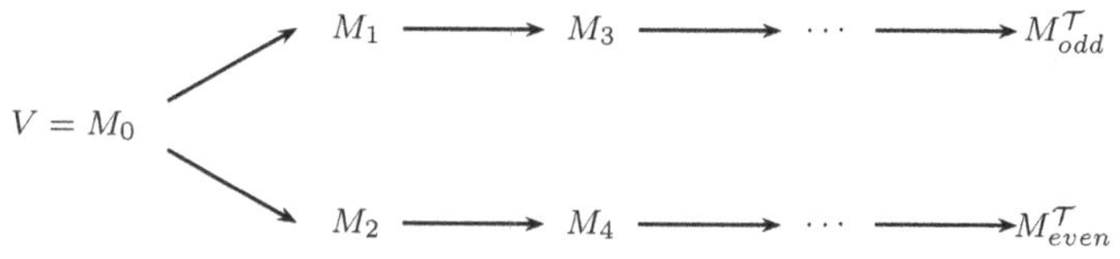

Fig. 20.

is a nice alternating chain on V, then either $M^{\mathcal{T}}_{odd}$ or $M^{\mathcal{T}}_{even}$ is well-founded. It is open whether one of the two must be ill-founded! More generally, we have

Big Open Problem. Is there a nice iteration tree on V of length ω having distinct cofinal well-founded branches?

A negative answer would mean that every such iteration tree can be continued in a **unique** way. This would have many useful consequences, as we shall see below.

Proof: [Proof of Theorem 7.10] Let $\mathcal{T} = \langle T, \langle M_n, E_n \mid n < \omega \rangle \rangle$ be a counterexample. The first step is to localize the bad-ness of $\mathcal{T}$ in countably many ordinals. (Compare Lemma 3.6.) This is done in the

Claim. There are ordinals α_n, for $n < \omega$, such that for all n, m

$$nTm \Longrightarrow i^{\mathcal{T}}_{nm}(\alpha_n) > \alpha_m.$$

Proof: [Proof of the Claim] First pick η such that whenever b is a cofinal barnch of T, then $i_b^{\mathcal{T}}(\eta)$ is in the ill-founded part of $M_b^{\mathcal{T}}$. (Here $i_b^{\mathcal{T}} : M_0^{\mathcal{T}} \to M_b^{\mathcal{T}}$ is the direct limit map.)

For $n < \omega$, let

$$\mathscr{B}_n = \{b \mid b \text{ is a confinal branch of } T \text{ and } n \in b\}.$$

Let

$$X = \{(n, f) \mid f : \mathscr{B}_n \to \eta\}.$$

For $(n, f), (m, g) \in X$, we let

$$(n, f) < (m, g) \quad \text{iff} \quad mTn \text{ and } \forall b \in \mathscr{B}_n(f(b) < g(b)).$$

It is easy to see that $<$ is a well-founded relation on X.

Now if $\mathscr{B}_n = \emptyset$, so that T is well-founded below n, we put

$$\alpha_n = |n|_T = \text{ rank of } T \text{ below } n.$$

Then if $\mathscr{B}_n = \emptyset$ and nTm, we have $\mathscr{B}_m = \emptyset$, and since $\alpha_n < \omega_1$,

$$i_{nm}(\alpha_n) = \alpha_n > \alpha_m,$$

as desired.

So we may assume $\mathscr{B}_0 \neq \emptyset$, otherwise we're done. For each $b \in \mathscr{B}_0$, pick α_n^b for $n < \omega$ such that

$$i_{nm}(\alpha_n^b) > \alpha_m^b$$

whenever nTm and $n, m \in b$. We may assume $\alpha_0^b = \eta$ for all b. Now for n such that $\mathscr{B}_n \neq \emptyset$, set

$$\alpha_n = \omega_1 + |(n, h)|_{i_{0,n}(<)}$$

where $h(b) = \alpha_n^b$ for all $b \in \mathscr{B}_n$. Note at this point that $h \in M_n$, since M_n is $2^{\aleph_0}$-closed. Then we have, if nTm and $\mathscr{B}_m \neq \emptyset$:

$$\begin{aligned} i_{n,m}(\alpha_n) &= \omega_1 + |(n, i_{nm}(\lambda b \in \mathscr{B}_n.\alpha_n^b))|_{i_{0m}(<)} \\ &= \omega_1 + |(n, \lambda b \in \mathscr{B}_n.i_{nm}(\alpha_n^b))|_{i_{0m}(<)} \\ &> \omega_1 + |(m, \lambda b \in \mathscr{B}_m.\alpha_m^b)|_{i_{0m}(<)} \\ &= \omega_1 + \alpha_m. \end{aligned}$$

This yields the claim. □

Ordinal s $\langle \alpha_n \mid n \in \omega \rangle$ as in the claim are said to witness that $\mathcal{T}$ is **continuously ill-founded**.

To simplify the rest of the proof a bit, we shall assume there are arbitrarily large ξ such that $V_\xi \models \mathbf{ZFC}$. We leave is as an exercise to dispense with this assumption.

Let $\langle \alpha_n \mid n < \omega \rangle$ witness that $\mathcal{T}$ is continuously ill-founded. Let $\eta_0 > \alpha_0$ be such that $V_{\eta_0} \models \mathbf{ZFC}$, and $\mathcal{T} \in V_\eta$. Now let N_0 be countable and transitive, and

$$\pi_0 : N_0 \to V_{\eta_0}$$

elementary, with $\mathcal{T} \in \operatorname{ran}(\pi_0)$, and $\langle \alpha_n \mid n < \omega \rangle \in \operatorname{ran}(\pi_0)$. Let

$$\pi(F_n) = E_n,$$

and

$$\pi(\beta_n) = \alpha_n$$

for all n. Note $\pi(T) = T$, and $\mathcal{U} = \langle T, \langle N_k, F_k \mid k < \omega \rangle \rangle$ is an iteration tree on N_0, where

$$N_{k+1} = \operatorname{Ult}(N_i, E_k), \text{ for } i = \operatorname{pd}_T(k+1).$$

Moreover, $\langle \beta_n \mid n < \omega \rangle$ witness that $\mathcal{U}$ is continuously ill-founded.

We now define $(P_m, \in, \eta_m)$ and π_m by induction on m so that

(1) $\pi_m : N_m \to V^{P_m}_{\eta_m}$,
(2) $P_m \models \mathbf{ZFC}$, P_m is closed under ω-sequences and $\left\{\alpha \mid \eta_m < \alpha < \operatorname{Ord}^{P_m} \wedge V^{P_m}_\alpha \models \mathbf{ZFC}\right\}$ has order-type $\geq \pi_m(\theta_m)$,
(3) for all $k \leq m$
  (a) $P_k \underset{\mathcal{T}}{\sim} P_m$, where $\mathcal{T} = \sup \pi_k"\operatorname{lh}(F_k)$,
  (b) $\pi_k \restriction V^{N_k}_{\operatorname{lh}(F_k)} = \pi_m \restriction V^{N_k}_{\operatorname{lh}(F_k)}$ and
(4) $P_m \in P_{m-1}$, if $m > 0$.

If we do this, clause (4) yields the desired contradiction.

$m = 0$: We have π_0 and η_0 already. Let $P_0 = V_\tau$, where $V_\tau \models \mathbf{ZFC}$ and $\{\alpha \mid \eta_0 < \alpha < \tau \wedge V_\alpha \models \mathbf{ZFC}\}$ has order type at least $\pi_0(\theta_0)$.

$m = n+1$: Let $E = \pi_n(F_n)$. Let $E = \pi_n(F_n)$. Let $\kappa = \operatorname{pd}_T(n+1)$, and $\tau = \sup \pi_k"\operatorname{lh}(F_k)$. $\operatorname{crit}(F_n) < \operatorname{lh}(F_k)$, so $\operatorname{crit}(E) < \tau$. Also, $P_m \sim_\tau P_k$. Hence we may set

$$Q = \operatorname{Ult}(P_k, E).$$

We have that Q is closed under ω-sequences, and hence well-founded. Let $j : P_k \to Q$ be the canonical embedding. Set

$$\gamma = j(\eta_k),$$

we can now find σ such that the diagram

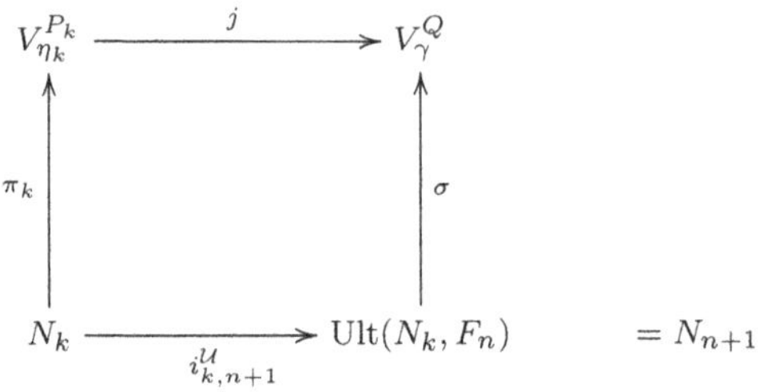

Fig. 21.

commutes. Namely, we set

$$\sigma([a, f]_{F_n}^{N_k}) = [\pi_n(a), \pi_k(f)]_E^{P_k}.$$

Exercise 33: Prove that σ is well-defined, elementary, and that the diagram commutes.

Exercise 34: Show that $\sigma \restriction V_{\mathrm{lh}(F_n)^{N_{n+1}}} = \pi_n \restriction V_{\mathrm{lh}(F_n)}^{N_n}$.

Together, these exercise are called the Shift Lemma. The key to proving them is that π_n and π_k agree on $V_{\mathrm{lh}(F_k)}^{N_k}$, by induction.

We could now set $P_{n+1} = Q$ and $\eta_{n+1} = \gamma$ and $\pi_{n+1} = \sigma$, and satisfy (1)-(3). In order to satisfy (4) as well, we replace Q by a Skolem hull of itself. Notice that Q has

$$j(\pi_k(\theta_k)) = \sigma(i_{k,n+1}(\theta_k)) > \sigma(\theta_{n+1})$$

many $\alpha > \gamma$ such that $V_\alpha^Q \models \mathbf{ZFC}$, in order type. Note also $\sigma \in Q$, as Q is ω-closed. Let

$$\mu = \sigma(\theta_{m+1})^{th} \text{ ordinal } \alpha > \gamma \text{ such that } V_\alpha^Q \models \mathbf{ZFC}.$$

Put

$$P_{n+1} = \text{ transitive collapse of } \mathrm{Hull}^{V_\mu^Q}(\{\sigma\} \cup V^Q_{\sup \sigma\text{"}\mathrm{lh}(F_n)}),$$

and

$$\pi_{n+1} = \text{ image of } \sigma \text{ under the collapse,}$$

$$\eta_{n+1} = \text{ image of } \gamma \text{ under the collapse.}$$

Notice that $P_{n+1} \in V^Q_{\mathrm{lh}(E)} = V^{P_n}_{\mathrm{lh}(E)}$, as $sup\sigma$"$\mathrm{lh}(F_n) < \mathrm{lh}(E)$, because $\mathrm{lh}(E)$ is inaccessible in Q. This gives (4). We leave (1)-(3) to the reader. □

8. Iteration Trees of Transfinite Length

It is often important to continue iterating into the transfinite. The way we continue a tree $\mathcal{T}$ of limit length λ, i.e.: pick a branch b which has been visited cofinally often below λ, and such that $\mathfrak{M}^{\mathcal{T}}_b$ is well-founded. Set $\mathfrak{M}^{\mathcal{T}}_\lambda = \mathfrak{M}^{\mathcal{T}}_b$, and continue. To be more precise:

Definition 8.1: Let $\gamma \in \mathrm{Ord}$. We call $\mathcal{T}$ a **tree order on** γ iff

(1) T is a strict partial order of γ,
(2) $\forall \beta < \gamma\ (T \text{ wellorders } \{\alpha \mid \alpha T \beta\})$,
(3) $\forall \alpha, \beta\ (\alpha T \beta \implies \alpha < \beta)$,
(4) $\forall \alpha\ (0 < \alpha \implies 0 T \alpha)$,
(5) $\forall \alpha$ (α is a successor ordinal iff α is a T-successor), and
(6) $\forall \lambda < \gamma$ (λ is a limit ordinal $\implies \{\alpha \mid \alpha T \lambda\}$ is $\in$-cofinal in λ).

Definition 8.2: Let $\gamma \in \mathrm{Ord}$. A **nice iteration tree of length** γ **on** M is a system

$$\mathcal{T} = \langle T, \langle (M_\alpha, E_\alpha) \mid \alpha < \gamma \rangle, \langle i_{\alpha\beta} \mid \alpha T \beta \rangle \rangle$$

such that

(1) T is a tree order on γ,
(2) $M_0 = M$,
(3) $M_\alpha \models E_\alpha$ is a nice extender,
(4) $\alpha < \beta \implies \mathrm{lh}(E_\alpha) < \mathrm{lh}(E_\beta)$,
(5) if $\alpha + 1 < \gamma$, then $M_{\alpha+1} = \mathrm{Ult}(M_\xi, E_\alpha)$, where ξ is least such that

$$M_\xi \;\overline{}_{\mathrm{crit}(E_\alpha)+1}\; M_\alpha.$$

(6) if $\lambda < \gamma$ is a limit ordinal, then

$$M_\lambda = \text{dir}\lim_{\alpha T \lambda} M_\alpha,$$
$$i_{\alpha\lambda} = \text{canonical embedding, for } \alpha T \lambda.$$

Is it always possible to continue a nice iteration tree on V? At successor steps, yes.

Theorem 8.3: *Let $M \models$ **ZFC** be transitive and closed under ω-sequences. Let $\mathcal{T}$ be a nice iteration tree on M of length $\alpha+1$, and let $\xi \le \alpha$ be such that $\mathfrak{M}^{\mathcal{T}}_\xi \xrightarrow[\text{crit}(E^{\mathcal{T}}_\alpha)+1]{} \mathfrak{M}^{\mathcal{T}}_\alpha$. Then $\text{Ult}(\mathfrak{M}^{\mathcal{T}}_\xi, E^{\mathcal{T}}_\alpha)$ is well-founded.*

Proof: We need the following exercise:

Exercise 35: Let $\mathcal{T}$ be a nice iteration tree and $\alpha < \text{lh}(\mathcal{T})$. Show

$$\mathfrak{M}^{\mathcal{T}}_\alpha = \{i^{\mathcal{T}}_{0\alpha}(f)(a) \mid f \in \mathfrak{M}^{\mathcal{T}}_0 \wedge a \in [\nu]^{<\omega}\}$$

where $\nu = \sup\{\text{lh}(E^{\mathcal{T}}_\xi) \mid (\xi+1)T\alpha \text{ or } \xi+1=\alpha\}$.

The exercise generalizes the fact that

$$\text{Ult}(M, E) = \{i^M_E(f)(a) \mid f \in M \wedge a \in [\text{lh}(E)]^{<\omega}\}.$$

It says that $\mathfrak{M}^{\mathcal{T}}_\alpha$ is Skolem-generated by $\text{ran}(i^{\mathcal{T}}_{0\alpha})$ together with ordinals below the sup of the lengths of extenders used on the branch 0-to-α.

Now let ξ, α be as in the theorem, and set $N = \text{Ult}(M^{\mathcal{T}}_\xi, E^{\mathcal{T}}_\alpha)$. Let $i : M \to N$ be the canonical embedding. ($i = \pi \circ i^{\mathcal{T}}_{0\xi}$, where $\pi : M^{\mathcal{T}}_\xi \to N$.) Let $\lambda = \text{lh}(E^{\mathcal{T}}_\alpha)$. The proof of Exercise 35 easily yields

$$N = \{i(f)(a) \mid a \in [\lambda]^{<\omega} \wedge f \in M\}.$$

Note here that although N may be ill-founded, $i(\kappa) \in \text{wfp}(N)$, where $\kappa = \text{crit}(E^{\mathcal{T}}_\alpha)$. This is because $V^N_{i(\kappa)+1} = V^{\text{Ult}(M^{\mathcal{T}}_\alpha, E^{\mathcal{T}}_\alpha)}_{i(\kappa)+1}$, and $V^{\text{Ult}(M^{\mathcal{T}}_\alpha, E^{\mathcal{T}}_\alpha)}_{i(\kappa)+1}$ can be computed in $M^{\mathcal{T}}_\alpha$, which thinks $E^{\mathcal{T}}_\alpha$ is an extender.

Now pick $\langle f_k \mid k < \omega\rangle$ such that there are $a_k \in [\lambda]^{<\omega}$ with $i(f_{k+1})(a_{k+1}) \in^N i(f_k)(a_k)$ for all k. Note $\langle f_k \mid k \in \omega\rangle \in M$! Thus

$$i(\langle f_k \mid k \in \omega\rangle) = \langle i(f_k) \mid k \in \omega\rangle \in N.$$

Now pick γ such that

$$N \models \gamma \in \text{Ord} \wedge \lambda < \gamma \wedge \langle i(f_k) \mid k \in \omega\rangle \in V_\gamma.$$

Working in N, we have H, π such that

$$N \models N \text{ is transitive, } |H| = \lambda, \text{ and } \pi : H \to V_\gamma$$
$$\text{with } \pi \restriction \lambda = id \text{ and } \langle i(f_k) \mid k \in \omega \rangle \in \mathrm{ran}(\pi).$$

Now $(\mathrm{ran}(\pi), \in^N)$ is ill-founded in V, hence $(H, \in^N)$ is ill-founded in V. But $H \in V^N_{i(\kappa)} \subseteq \mathrm{wfp}(N)$, a contradiction. □

Insofar as continuing nice trees on V at limit steps goes, the main result is the following

Theorem 8.4: *Let $\mathcal{T}$ be a nice iteration tree on V of countable limit length λ. Suppose that for all limit $\eta < \lambda$, $\{\alpha \mid \alpha T \eta\}$ is the* unique *cofinal well-founded branch of $\mathcal{T} \restriction \eta$. Then $\mathcal{T}$ has a cofinal, well-founded branch.*

Remark 8.5: In other words, if $\mathcal{T}$ has made the only choice it could make at limit $\eta < \lambda$, then there is a choice for it to make at λ.

The proof of Theorem 8.4 is much like the proof of Theorem 7.10 we gave above.

This leads us to one of the biggest open problems in the subject.

Definition 8.6: **Nice-UBH** is the statement: Every nice iteration tree on V of limit length has at most one cofinal, well-founded branch.

Definition 8.7: **Generic-nice-UBH** is the statement: $V[G] \models$ nice-UBH, whenever G is set generic over V.

Whether nice-UBH, or better generic-nice-UBH, are true are very important questions. Of course, the more useful answer would be "yes". The reason is that we would then get, via Theorem 8.4, an *iteration strategy* for V. We now explain that concept more precisely.

Let $M \models$ **ZFC** be transitive, and $\theta \in \mathrm{Ord}$. The (nice) **iteration game of length θ on M** is played as follows: there are two players, I and II. They cooperate to produce an iteration tree $\mathcal{T}$ on $\mathfrak{M}$. At successor rounds $\alpha + 1$, player I extends $\mathcal{T}$ by picking a nice $E^{\mathcal{T}}_\alpha$ from $\mathfrak{M}^{\mathcal{T}}_\alpha$, and setting $\mathfrak{M}^{\mathcal{T}}_{\alpha+1} = \mathrm{Ult}(M^{\mathcal{T}}_\xi, E^{\mathcal{T}}_\alpha)$ for ξ least such that $\mathrm{crit}(E^{\mathcal{T}}_\alpha) < \mathrm{lh}(E^{\mathcal{T}}_\xi)$. (If the ultrapower is ill-founded, the game ends, and I has won.) At limit rounds $\lambda < \theta$, II extends $\mathcal{T}$ by picking b cofinal in λ such that $\mathfrak{M}^{\mathcal{T}}_b$ is well-founded. If II fails to do this, I wins. If after θ rounds, I has not yet won, then II wins.

We call this game $G_{\text{nice}}(M, \theta)$. A winning strategy for II in $G_{\text{nice}}(M, \theta)$ is called a **θ-iteration strategy** for M. We say M **is θ-iterable (for nice trees)** iff there is a θ-iteration strategy for M. We have

Theorem 8.8: *If nice-UBH holds, then V is ω_1-iterable for nice trees.*

Proof: Player II's strategy in $G_{\text{nice}}(V, \omega_1)$ is: at round λ, pick the unique cofinal well-founded branch of $\mathcal{T} \restriction \lambda$. □

$\omega_1 + 1$-iterability is much more useful than ω_1-iterability. We have

Theorem 8.9: *If generic-nice-UBH holds, then V is κ-iterable for nice trees, where κ is the least measurable cardinal.*

Exercise 36: Prove Theorem 8.9. [You need to know something about preservation of large cardinals under small forcing to do this one, so it is really only for the more advanced students.]

The main result known in the direction of proving UBH is the following. For $\mathcal{T}$ a nice tree, set

$$\delta(\mathcal{T}) = \sup\{\text{lh}(E_\alpha^{\mathcal{T}}) \mid \alpha < \text{lh}(\mathcal{T})\},$$
$$\mathfrak{M}(\mathcal{T}) = \bigcup_{\alpha < \text{lh}(\mathcal{T})} V_{\text{lh}(E_\alpha^{\mathcal{T}})}^{M_\alpha^{\mathcal{T}}}.$$

So $\text{Ord} \cap \mathfrak{M}(\mathcal{T}) = \delta(\mathcal{T})$, and $\mathfrak{M}(\mathcal{T}) = V_{\delta(\mathcal{T})}^{M_b^{\mathcal{T}}}$ for any cofinal branch b of $\mathcal{T}$ such that $\delta(\mathcal{T}) \in M_b^{\mathcal{T}}$.

Theorem 8.10: *Let $\mathcal{T}$ be a nice iteration tree of limit length, $\delta = \delta(\mathcal{T})$, and suppose b and c are distinct cofinal branches of $\mathcal{T}$ such that $\delta \in M_b^{\mathcal{T}} \cap M_c^{\mathcal{T}}$. Let $A \subseteq \delta$ and $A \in M_b^{\mathcal{T}} \cap M_c^{\mathcal{T}}$. Then*

$$(\mathfrak{M}(\mathcal{T}), \in, A) \models \text{“}\exists\kappa\ (\kappa \textit{ is } A\textit{-reflecting in } \text{Ord})\text{”}.$$

Remark 8.11: Another way this is often stated is: $\delta(\mathcal{T})$ is Woodin with respect to all $A \in M_b^{\mathcal{T}} \cap M_c^{\mathcal{T}}$, with respect to extenders in $\mathfrak{M}(\mathcal{T})$.

Proof: [Proof of Theorem 8.10] (Sketch). Let us consider the special case $\mathcal{T}$ is an alternating chain, and b is its even branch and c is its odd branch. Note that the extenders of $\mathcal{T}$ overlap in the following "zipper" pattern.

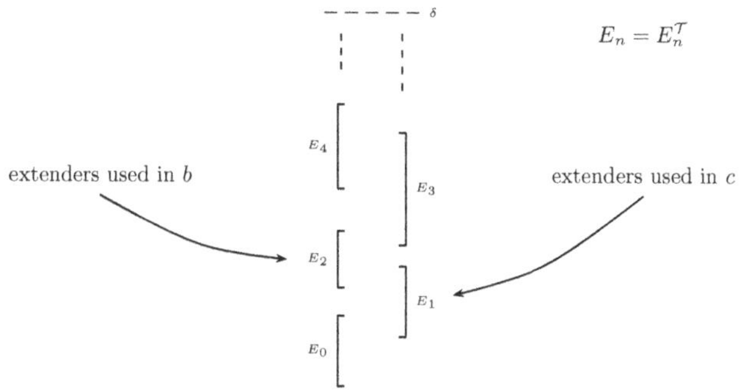

Fig. 22.

That is, letting $\kappa_n = \text{crit}(E_n)$ and $\lambda_n = \text{lh}(E_n)$: $\kappa_n < \kappa_{n+1} < \lambda_n$ for all n. Now let $A \subseteq \delta$ and $A \in M_b^{\mathcal{T}} \cap M_c^{\mathcal{T}}$. Pick m large enough that $A \in \text{ran}(i_{m,b}) \cap \text{ran}(i_{m+1,c})$.

Claim 2: For any $n \geq m$, $i_{E_n}(A \cap \kappa_n) \cap \lambda_n = A \cap \lambda_n$.

(That is, i_{E_n} shifts A to itself below the next critical points. It does not matter whether we write $i_{E_n}^{\mathfrak{M}(\mathcal{T})}$ or $i_{E_n}^{M_{n-1}^{\mathcal{T}}}$ here, since the ultrapowers are the same below the image of κ_n.)

Proof: [Proof of Claim] Suppose e.g. $n+1 \in b$. Let $A = i_{m,b}(\bar{A})$. Then $i_{m,n-1}(\bar{A}) \cap \kappa_n = A \cap \kappa_n$, because $\text{crit}(i_{n+1,b}) = \kappa_n$. So $i_{n-1,b}(A \cap \kappa_n)$ agrees with A below $i_{n-1,b}(A \cap \kappa_n)$. But $i_{n-1,b}(A \cap \kappa_n)$ agrees with $i_{E_n}(A \cap \kappa_n)$ below λ_n, because $\text{crit}(i_{n+1,b}) \geq \lambda_n$. Consider

$$i_{E_n \restriction \kappa_{n+1}} \circ \cdots \circ i_{E_{m+1} \restriction \kappa_{m+2}} \circ \cdots \circ i_{E_m \restriction \kappa_{m+1}} = j.$$

Note $E_i \restriction \kappa_{i+1} \in M(\mathcal{T})$, because $\kappa_{i+1} < \text{lh}(E_i)$! It is routine to show that j witnesses κ_m is A-reflecting to β in $M(\mathcal{T})$. □

Exercise 37:

(1) Complete the proof of this.
(2) Complete the proof of Theorem 8.10 as follows: let b and c be distinct cofinal branches of $\mathcal{T}$. Find $\langle \alpha_n \mid n \in \omega \rangle$ cofinal in λ such that for all n

$$\alpha_{2n} + 1 \in b, \alpha_{2n+1} + 1 \in c$$

and

$$\mathrm{crit}(E_{\alpha_k}) < \mathrm{crit}(E_{\alpha_{k+1}}) < \mathrm{lh}(E_{\alpha_k})$$

for all k. I.e. we have the zipper pattern embedded in the two branches. Now argue as above. □

Remark 8.12: For more detail, see [4].

Remark 8.13: So if $\mathcal{T}$ has distinct cofinal branches, then $\mathrm{lh}(\mathcal{T})$ has cofinality ω. This is also easy to see from the fact that every branch of an iteration tree is *closed* below its sup (as a set of ordinals).

Corollary 8.14: *Suppose nice-UBH fails. Then there is a proper class model with a Woodin cardinal.*

Proof: Let $\mathcal{T}$ be nice on V, and have distinct cofinal well-founded branches b and c. Then $L(M(\mathcal{T})) \subseteq M_b^{\mathcal{T}} \cap M_c^{\mathcal{T}}$. So by Theorem 8.10,

$$L(M(\mathcal{T})) \models \delta(\mathcal{T}) \text{ is Woodin.}$$ □

References

1. A. Andretta, "Building iteration trees", *Jour. Symb. Logic* **56**(1991), 1369–1384.
2. T. Jech, *Set Theory*, 3rd millennium edition, Springer-Verlag, Berlin, Heidelberg, New York, 2002.
3. K. Kunen and J. B. Paris, "Boolean extensions and measurable cardinals", *Ann. Math. Logic.* **2**(1970/71), no. 4, 359–377.
4. D. A. Martin and J. R. Steel, "Iteration trees", *J. Amer. Math. Soc.* **7**(1994), no. 1, 1–73.
5. W. J. Mitchell, "Sets constructed from sequences of ultrafilters", *Jour. Symb. Logic* **39**(1974), 57–66.
6. W. J. Mitchell, "Beginning inner model theory", *Handbook of Set Theory*, Springer-Verlag, Berlin, Heidelberg, New York, 2010, 1449–1495.

Printed in the United States
By Bookmasters